Practical Implementation of a Data Lake

Translating Customer Expectations into Tangible Technical Goals

Nayanjyoti Paul

Apress®

Practical Implementation of a Data Lake: Translating Customer Expectations into Tangible Technical Goals

Nayanjyoti Paul
Edison, NJ, USA

ISBN-13 (pbk): 978-1-4842-9734-6   ISBN-13 (electronic): 978-1-4842-9735-3
https://doi.org/10.1007/978-1-4842-9735-3

Managing Director, Apress Media LLC: Welmoed Spahr
Acquisitions Editor: Celestin Suresh John
Development Editor: James Markham
Coordinating Editor: Mark Powers

Cover designed by eStudioCalamar
Cover image by Arek Socha on Pixabay (www.pixabay.com)

Distributed to the book trade worldwide by Apress Media, LLC, 1 New York Plaza, New York, NY 10004, U.S.A. Phone 1-800-SPRINGER, fax (201) 348-4505, e-mail orders-ny@springer-sbm.com, or visit www.springeronline.com. Apress Media, LLC is a California LLC and the sole member (owner) is Springer Science + Business Media Finance Inc (SSBM Finance Inc). SSBM Finance Inc is a **Delaware** corporation.

For information on translations, please e-mail booktranslations@springernature.com; for reprint, paperback, or audio rights, please e-mail bookpermissions@springernature.com.

Apress titles may be purchased in bulk for academic, corporate, or promotional use. eBook versions and licenses are also available for most titles. For more information, reference our Print and eBook Bulk Sales web page at http://www.apress.com/bulk-sales.

Any source code or other supplementary material referenced by the author in this book is available to readers on GitHub (https://github.com/Apress). For more detailed information, please visit https://www.apress.com/gp/services/source-code.

Paper in this product is recyclable.

Table of Contents

About the Author

 **Nayanjyoti Paul** is an associate director and chief Azure architect for GenAI and LLM CoE for Accenture. He is the product owner and creator of patented assets. Presently, he leads multiple projects as a lead architect around generative AI, large language models, data analytics, and machine learning. Nayan is a certified master technology architect, certified data scientist, and certified Databricks champion with additional AWS and Azure certifications. He has been a speaker at conferences like Strata Conference, Data Works Summit, and AWS Reinvent. He also delivers guest lectures at universities.

About the Technical Reviewer

Arunkumar is an architect with 20+ years of experience in the IT industry. He has worked with a wide variety of technologies in the data, cloud, and AI spaces. He has experience working in a variety of industries such as banking, telecom, healthcare, and avionics. As a lifelong learner, he enjoys taking on new fields of study and challenging himself to master the necessary skills and knowledge.

Preface

This book explains how to implement a data lake strategy, covering the technical and business challenges architects commonly face. It also illustrates how and why client requirements should drive architectural decisions.

Drawing upon a specific case from my own experience, I begin with the consideration from which all subsequent decisions should flow: what does your customer need?

I also describe the importance of identifying key stakeholders and the key points to focus on when starting a project. Next, I take you through the business and technical requirements-gathering process and how to translate customer expectations into tangible technical goals.

From there, you'll gain insight into the security model that will allow you to establish security and legal guardrails, as well as different aspects of security from the end user's perspective. You'll learn which organizational roles need to be onboarded into the data lake, their responsibilities, the services they need access to, and how the hierarchy of escalations should work.

Subsequent chapters explore how to divide your data lakes into zones, organize data for security and access, manage data sensitivity, and use techniques for data obfuscation. Audit and logging capabilities in the data lake are also covered before a deep dive into designing data lakes to handle multiple file formats and access patterns. The book concludes by focusing on production operationalization and solutions to implement a production setup.

After completing this book, you will understand how to implement a data lake and the best practices to employ while doing so, and you will be armed with practical tips to solve business problems.

What You Will Learn

Specifically, by reading this book, you will

- Understand the challenges associated with implementing a data lake

- Explore the architectural patterns and processes used to design a new data lake

- Design and implement data lake capabilities

- Associate business requirements with technical deliverables to drive success

Who This Book Is For

This book was written for data scientists and architects, machine learning engineers, and software engineers.

Introduction

I landed at the airport and took an Uber to my customer's office. I was supposed to meet with the program manager on the customer side. After the initial process and getting myself "checked in," I entered the conference room that was booked for our team. I knew most of the team from other projects, but I was meeting a few of them for the first time. After the usual greetings and a few of my colleagues congratulating me on my new role, I was ready for the day to unfold.

This customer was a big organization, and there was a clear "separation of concerns" from multiple teams. The schedule was set up, and our first tasks were to get acquainted with the different organizational units, identify the key stakeholders, and understand the stakeholders' primary "asks." It was important for my team to understand the key organizational units and have one-on-one initial discussions. We needed to connect with the following people and teams:

- We needed to know the owner of this platform. This typically includes who will own this data lake as a platform from the customer's point of view. Who will pay the bills and eventually be the key decision-maker for all technical and business decision-making? We identified the senior VP of engineering as the key stakeholder. We set up a one-hour call with him to understand his expectations and his vision of the future-state data lake.

- We wanted to know the team that was handling all the data and analytics today. As the customer had an on-premise footprint, we wanted to know the engineering team who had been managing the entire data and analytics platform on-premise up to now. Eventually they would be cross-trained and be the data engineering team in the cloud after we delivered the data lake. As all the information of source systems, data onboarding processes, current business reporting needs, etc., were managed by them, we needed to understand the current business process of this team and document them so that we could draw some parallels for what it might take to transition those workload and business requirements into the cloud as part of this journey. We invited the engineering leads to an initial one-hour call.

- We needed to connect with the chief information security officer (CISO) and her team. Venturing into the cloud was a new entity for my customer. Apart from the technical questions and recommendations, we needed to understand the business, contractual, and general organizational obligations of what was permitted (and what was not) from a security standpoint. We knew that every organization has a set of predefined policies that must be followed. Some of these guidelines come from geography (like GDPR), some come from industry (like HIPAA or financial data restrictions), and others may come from data residency (like data sitting in the customer's own on-premise data center versus the public cloud). Nevertheless, we needed to connect with this team and understand what these policies meant for this customer and what considerations we needed to take when we designing the platform as a whole. We ended up setting up another one-hour call with this team.

- Next we set up a call with the "cloud engineering" team. This was a new team, and they had started some groundwork in laying out the "laws of the land," mostly in terms of network, services whitelisted, getting access to a cloud network, access and onboarding of resources to the cloud system, etc. We wanted to be acquainted with the current process. Also, from a delivery point of view, this project was a shared responsibility. Some of the key aspects that our customer would still be "owning" was the platform management and onboarding part. Additionally, the strategies around disaster recovery, high availability, etc., were going to be a "shared responsibility." Hence, it was critical for us to work closely with the cloud engineering team, so we scheduled a one-hour initial discussion with them.

- Next was the DBA team. The DBA team currently owned the databases on-premise but was also responsible for eventually owning any databases, data marts, and data warehouses that would be set up on the cloud as part of this program. We set up a one-hour meeting with them too.

- Next was the data governance team. One of the key reasons to move into the cloud (apart from the obvious reasons of low-lost, easy maintenance, and limitless storage and compute capacity) was to keep track of and audit everything that was going on. We believed in a "governance-first" approach, and our customer believed in that too. They wanted to keep an audit and lineage trail of everything that would be happening on the cloud so that the data lake (lake house) did not become a swamp. An easy and centralized governance process would make "things" in the data lake very organized. Additionally, it would introduce data discovery and search capability that would become a crucial feature for building and establishing a data marketplace and catalog to "shop for" all the data (products) hosted on the data lake (lake house).

– We also connected with the "business" users who were the key stakeholders of the system. They were supposed to use and consume data or analytics outcomes from the platform. We had teams like data science, business intelligence, C-suite executives, etc., who were waiting to be onboarded onto the platform for different reasons and rationales. We set up independent calls with them to understand what "success" meant for them.

– Lastly, we wanted to quickly connect with our partner teams. For example, the public cloud offering was from AWS, and we wanted to connect with the AWS leads to understand what was currently in discussion for this implementation. Similarly, we connected with the Collibra team that was providing the Collibra software as an enterprise data catalog solution. Coming from a consulting company, we have partnerships with both vendors, and hence it was critical for us to be in sync with them.

With the key stakeholders identified and meetings set up, it was time for business. Having dedicated sessions with each key member was critical to get "buy-in" from each of them for the platform architecture (more on this to follow in the coming chapters).

Understanding the Requirements from Multiple Stakeholders' Viewpoints

In general, implementing a greenfield data lake has many technical and business challenges. The following are a few challenges that we needed to think through:

- Establishing a clear understanding of the customer requirements for a data lake implementation can be a challenge because of the complexity of the area.

- It can be difficult to determine exactly what data is required, as well as how it should be stored and retrieved.

- It is difficult to understand the customer's desired outcomes and how they will use the data lake.

- It can be challenging to ensure that the data lake is secure and conforms to industry standards and regulations.

- Connecting the data lake with other systems can be a challenge because of the complexity of the integration.

- It can be difficult to determine the best way to structure the data lake, as well as how to optimize it for performance.

- It is difficult to ensure that the data lake is designed for scalability so that it can accommodate future growth.

- Determining the most effective way to ingest data into the data lake can be a challenge because of the volume and variety of data sources.

- It can be difficult to ensure that the data is of high quality, as well as how to monitor and maintain the data within the data lake.

- Since the customer requirements will vary from one organization to the next, it can be difficult to have an accurate understanding of what is needed and build a generalized solution.

- Understanding the customer's security and privacy requirements can be difficult to interpret, especially if they are not adequately documented.

- Establishing the necessary data governance frameworks and policies can be a challenge if there is not sufficient detail regarding the customer's requirements.

- Understanding the customer's desired access and usage policies can be difficult to discern without an appropriate level of detail in the customer's requirements.

- Establishing the necessary data quality requirements can be a challenge if the customer's requirements are not met.

The following diagram represents how "success" means different things to different stakeholders. This illustration depicts an example of what it means for this particular customer. This is to ensure that we address and keep each of these success criterion in mind as we move ahead and start the platform design.

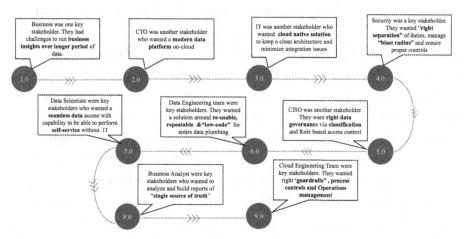

If we look closely, the first stakeholders are from the business side. For them, the objective is outcome focused. The technology is secondary for them as long as we continue delivering high-quality business insights in a repeatable and predictable time frame.

Second are the stakeholders from the CTO's office. They want to design the platform (data lake) as a future-ready solution. For them it is important to make the right technical decisions and adopt a cloud-first approach. They want to focus on a modern data stack that centers around cloud-native and software-as-a-service (SaaS) offerings.

Next, the customer's IT organization is a key stakeholder. Their focus is to incorporate technical solutions that are easy to maintain, cloud native, and based on the principles of keeping the integrations minimal.

Next in line as a key stakeholder is the security office team. They want to ensure that we design a system that has the right "separation of concerns" and has the right security guardrails so that confidential and personally identifiable information (PII) data can be safe and secure.

Next in line is the CISO's team for whom the data access policies, data governance and auditability, etc., are primary concerns. They want to ensure that the data is available only to the right resources at the right time through role-, tag-, and attribute-based access controls.

Next in line is the data engineering team who will eventually "own" the applications and system for maintenance. For them it was important that the data engineering solution built on the data lake has reusability, extensibility, and customizability, and is based on a solid programming framework and design that will be easy to manage and use in the long run.

Next in line is the data scientist community who needs the right access to the data and right access to the tools to convert the data into insights. They also want "self-service" as a capability where they have the right permissions to work on ideas that can help the business get value.

Next in line is the business analyst community who want to be onboarded into this new data lake platform as soon as possible with access to a "single source of truth" so that they can start building the mission-critical application that the business is waiting for.

Finally, the cloud engineering team is a key stakeholder. This team wants the whole platform to be secure, controlled, user friendly, reliable, and durable.

As you might have imagined by now, I will be using my experience to explain the end-to-end process of designing and implementing a data lake strategy in the following chapters.

This book will (in broad strokes) cover concepts such as how to understand and document the business asks, define the security model, define the organization structure, design and implement the data lake from end to end, set up a production playground, and operationalize the data lake. Finally, I will present some lessons learned from my experience.

Chapter 1 will focus on each of these points and how each resulted in the design of a small part of the key problem (platform design) and how little by little things fell into place for me and my team. Let's get started.

Understanding "the Ask"

Objective: Asking the Right Questions

In the introduction of the book, I set the stage for the project we'll start discussing in this chapter. When I took up the solution architect and delivery lead role, I had no idea what vision my customer had, other than a very general understanding of the final product my customer was after. The intention was to build a modern, cloud-centric data and analytics platform (called a *lake house*). So, at this point, it was important for me and my team to ask the right questions, gather the requirements in detail, and start peeling back the layers of the onion. In short, we needed to understand "the ask."

The first ask (for my team and me) was to be aligned to the customer's vision. To understand this vision, we set up a meeting with the VP of engineering (the platform owner) to establish the direction of the project and the key decisions that needed to be made.

© Nayanjyoti Paul 2023
N. Paul, *Practical Implementation of a Data Lake*,
https://doi.org/10.1007/978-1-4842-9735-3_1

The Recommendations

I used the following checklist as part of the vision alignment, and you can use this for your project too. Also, be open to bringing your own questions to the meeting based on your customer's interests and their maturity.

- What are the migration path, modernization techniques, enhancements, and cloud vendor that will be used?

- What are the current challenges?

- Why is modernizing data platforms hard?

- What are the top five issues that we want to solve?

- What is available on-premise and on the cloud already?

- What meetings will be needed throughout the project?

- What common terms and jargon can we define?

My team and I started the first round of discussions with the key customer stakeholders. We then understood the requirements better and had a better appreciation of the direction our customer wanted to go in. Each of the seven topics listed previously are detailed in the remainder of the chapter.

Decide on the Migration Path, Modernization Techniques, Enhancements, and the Cloud Vendor

After the usual greetings and formal introduction, we sat down to start documenting the vision. We understood that the requirement was to build a cloud-native and future-proof data and analytics platform. Having said that, the high-level objective was very clear. The data lake design was supposed to be sponsored by the business, and they had strict timelines to ensure we could get 25 highly important reports ready. Of the 25 reports, most of them were to be built on business logic after bringing in data from the system of records, but a few of them were to be powered by machine learning predictive models. For us, that meant that the business had a very specific "success criteria" in mind, and as long as we could deliver on the business promise (through a technical capability), we could deliver value.

Even though the outcome was business focused, the enabler was technology. We wanted to design the architecture "right" so that we could have a sustainable and adaptive platform for data and analytics for the future.

We started asking specific questions focused on whether the customer had already chosen a cloud partner. This was critical as we wanted to be cloud-native and leverage the capabilities each cloud vendor provided. In this case, the customer already had decided on AWS. Questions around whether the requirement was to modernize an existing architecture, migrate a similar technology, or enhance an existing setup were important for us to understand. Table 1-1 provides a quick reference for each question we asked and why it was important.

These questions can add value to any project during the initial understanding phase. Feel free to use Table 1-1 as a baseline for documenting the basic premises of the offering you are planning to deliver for your customer.

Table 1-1. *Assessment Questions*

Questions	Why Was the Question Important?	What Was Decided?
What cloud platform to use?	Each year cloud vendors introduce new capabilities, features, and integrations. By being aligned to a cloud vendor's capabilities, we can understand the "out-of-box" offerings versus gaps for that specific vendor. Also this means a correct estimation for time and cost based on the maturity of the vendor and the capabilities they currently offer.	The customer's decision to go with AWS ensured (for example) that we could leverage its ML capabilities on Sagemaker, their centralized RBAC and TBAC policies through lake formation, and many more (more on those later).

(continued)

Table 1-1. (*continued*)

Questions	Why Was the Question Important?	What Was Decided?
Do you want to implement a lift-and-shift, modernization, or migration solution strategy?	Each of these solutions needs separate handling and enablement from a technical point of view. For example, lift and shift should focus on a path of least resistance to have the same capability available in the cloud. So, an Oracle system on-premise can be deployed as an Oracle system on the cloud. Migration is slightly different; for example, the same Oracle system can be migrated to a Redshift system on the cloud leveraging native cloud capabilities but keeping the basics intact. However, modernization can mean replacing an on-premise system like Oracle with a data lake or a lake house architecture where we can enable different personas such as data engineers, analysts, BI team, and the data science team to leverage the data in different ways and with different forms to get value.	The customer was very clear that they wanted a data lake in the cloud, which meant they were ready to open up new possibilities, new personas, new kinds of use cases, and new opportunities for the whole organization.

Assess the Current Challenges

Even though the vision from the customer was for a new, modern data platform, it is always important to understand why the customer has decided to take that initiative now, including what challenges have become important enough that they could not sustain the existing solution. Also, documenting their current challenges provides a great way to evaluate "success" and measure the outcomes. The following were some of the critical challenges that were high priority for our customer in this example:

- The current setup was costly. The software vendors for the commercial off-the-shelf (COTS) products were charging a license fee based on the number of machines. As the organization was growing, so was their user base.

- The current setup could not scale up based on the organization's needs, seasonality, user personas, etc.

- As the data volume was growing, the current trend of analytics was very slow and restrictive. There was no option for machine learning, predictive modeling, unstructured data analysis, etc.

- As the organization was gearing up for the future, they had started investing in data scientists, data analysts, etc. The organization had started recruiting new talent, and it was important to build a platform that helped them bring in value.

- Time to market was essential, and a process that can provide "self-service" capabilities and quick prototyping features can unlock a lot of capabilities for the customer.

- They wanted to be future ready. Peer pressure is a huge motivation. As other organizations in the same space were adapting to the world of cloud-native and cloud-centric solutions, it was important for our customer to not fall behind.

Some of these points were critical for the customer, and hence we ensured that when we designed the solution, we considered the people who would be using the platform and what capabilities the final platform should have.

Understand Why Modernizing Data Platforms Is Hard

Along with identifying the customer's challenges and issues that they were currently facing, it was important to have an open discussion on the challenges other customers have faced (in similar domains) and what we had learned through our experiences (lessons learned). Figure 1-1 provides a quick reference for our past experience, which we thought would help this current customer to see a pattern and help us avoid common gotchas.

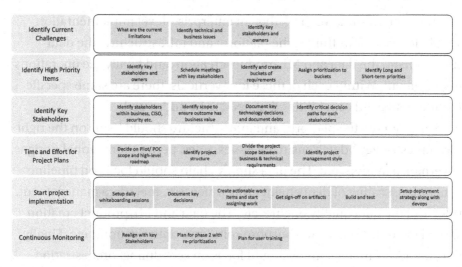

Figure 1-1. *High-level step-by-step process of organizing the project through different phases*

Along with the Figure 2-1 pointers on what we should focus on while delivering an enterprise-scale data platform solution, Figure 1-2 provides guidelines for a target-state implementation as part of an end-to-end data platform implementation.

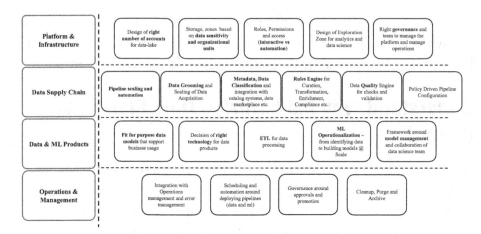

Figure 1-2. *Holistic view of end-to -end capabilities needed for a data strategy project (from a technical view)*

At the minimum, we think that an enterprise-level implementation should focus on four things: the platform and infrastructure, the data supply chain, the data and ML product creation, and finally the operations management. Within each of these four verticals, there are some specific requirements and capabilities that need to be addressed.

For example, the platform and infrastructure should focus on the right "account strategy," data sensitivity issues, roles, permissions, zonal designs, governance models, etc. The data supply chain should focus on pipeline generation, scaling of data engineering processes, metadata management, rules and the data quality engine, etc. The data and ML product creation should focus on ETL, fit-for-purpose solutions, optimizations, etc. Finally, the operations management should focus on scheduling, orchestration, etc. Figure 1-2 is just a representation, but it provides a blueprint in ensuring that we think through all these capabilities while designing and implementing the end-to-end enterprise data platform.

Determine the Top Five Issues to Solve

This is similar to the previous point discussed. However, the key differentiation is the process of collecting this information. To understand the top five issues, we started interviewing almost 50+ key associates and documenting the top issues they faced. We collected and collated the answers to our questions across different organization units based on where the key stakeholders were aligned. The result of the interview process was a list of common issues faced across the organization. When we looked at the report, we found many repetitive and common challenges. Those challenges surely impacted a lot of business units and hence were high on the priority list for us. Here are a few examples of the common challenges:

- The customer needed a central data hub. Different business units had independent silos of data repositories, which were either stale or out of sync.

- There was no single 360-degree view of data. Every business unit could see only their own side of data.

- Governance was nonexistent. There was no central catalog view of organization-wide datasets.

- Time to market was slow because of the limitations of the technology.

- The cost of management and maintenance was a fundamental issue.

What we discovered from this process was aligned to what we expected from the previous step, but seeing a repetitive pattern gave us the confidence that we had been following the right path so far and documenting the right issues. Figure 1-3 gives a pictorial view of this.

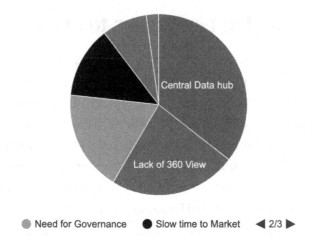

Figure 1-3. *A pie chart of what we saw were the driving factors for the need to build a data lake solution*

Determine What Is Available On-Premise vs. on the Cloud

It is important to understand our current situation and assess customer maturity before committing and undertaking any journey. Next, what we did with our key stakeholders was to understand from them where they stood and assess where they were currently in their vision. This would help us to offer the right help and guidance to reach the goal.

First, we sat down with the cloud security and infrastructure team to understand if the customer had started any journey toward AWS (their chosen platform). Next, we wanted to understand if any guidelines, corporate policies, and/or best practices were documented. Table 1-2 summarizes what details we got from the team. (Use this as a guide for your project, but ensure you have a customer document these, as they will become the rules of the game.)

Table 1-2. *Maturity Assessment Questionnaire*

Questions	Maturity Assessment
Has the organization started the journey toward the cloud in practice, or is it still a "paper exercise"?	In this case, the customer had a well-established cloud engineering practice. However, they had not implemented any large-scale implementation in the cloud. It had only a few small proofs of concept for a smaller group within the organization.
Does the organization have any standard security or cloud practices documented already?	The customer had documentation around cloud policies and best practices. However, the customer wanted us to review them, find gaps, and propose a better approach for the future.
Who are the personas (teams) with access to the current on-premise data warehouse the customers are hosting? Is the intention of the customer to onboard other personas in the new data platform (when ready), and will this imply a different set of access policies and practices?	The customer wanted the platform to be built to be future proof and ready for other organizational units to feel secure enough with it to onboard their analytics workload. This meant that we had to think beyond what the current on-premise systems provided in terms of role-based, attribute-based, and domain-based access to data and build a solution that would provide a separation of concerns for each team who would use the platform and onboard their data for analytics.

(*continued*)

Table 1-2. (*continued*)

Questions	Maturity Assessment
Have the consumption patterns changed? Are there new parties and use cases that would be adopted on the new platform?	The simple answer was yes. A major focus was to onboard the data science teams and enable them to build cutting-edge use cases to help do predictive insights on data rather than reactive ones. Similarly, a new kind of data analytics and BI teams would need instant and live access to the data to build and refresh metrics for the business to help in quick decision-making. Those personas and their set of use cases were completely new and unknown and would surely need a different design approach.
Do you want to be provider agnostic or multicloud (from a strategy point)?	Most customers start with the idea of setting up a cloud-based system targeting a specific cloud provider for partnership. However, soon clients decide to have a multicloud strategy that is provider agnostic. These decisions do not impact the solution strategy in the short to medium run, but they do have implications in the long run. For this customer, they did not have any preference about this, and we were supposed to focus on the AWS-specific solution for now.

Create the Meetings Needed Throughout the Project

Implementing a large-scale project is always challenging. Typically when we have sprint-based programs and each sprint is 2 weeks, it is important to think ahead and plan for the upcoming tasks. So, we wanted to identify important meetings and get them on the calendar. This included identifying the priority and ordering of tasks and ensuring we got calendar time from each stakeholder so that we did not have to wait for any important decisions from our customers.

We enabled three workstreams. I ensured we had dedicated teams for each of the three workstreams, and each had specific responsibility areas, as listed in Table 1-3. You can use this table to plan ahead for important meetings with the right stakeholders.

***Table 1-3.** High-Level Workstreams with Their Typical Responsibilities for a Technical Data Lake Implementation*

Workstream	Main Responsibilities
Business analysis and grooming	— Identify and prioritize source systems that need to be onboarded into the new platform.
	— Identify which datasets from which sources need to be priority 1.
	— For each source, "groom" the dataset based on data profile, types of data, type of interactions, frequency of loads, and special data handling needs (version of data versus snapshot, etc.).
	— For each dataset, document basic data quality checks, common issues, common standardization needs, and common enrichment needs required.
	— From a consumption point of view, clearly document the ask, expected business outcome, and samples of output.
	— From a consumption point of view, clearly document the business logic for converting source datasets into the expected outcome.

(continued)

Table 1-3. (*continued*)

Workstream	Main Responsibilities
Data security	— Work with the data security teams, CISO teams, and cloud engineering teams, and have a common understanding of how many AWS accounts are needed, how many environments are needed (dev/UAT/prod), how to separate out the concerns of "blast radius," how to manage data encryption, how to manage PII data, how to implement network security on data onboarding and IAM policies, etc.
	— Identify and document processes to define how to onboard a new source system and what access and security should be in place.
	— Identify and document processes to define a user onboarding process through AD integrations, IAM policies, and roles to be applied.
	— Have separate capabilities between interactive access and automated access and have different policies, services, and guardrails for both types.
	— Understand and document life-cycle policies and practices for data and processes.
	— Understand and document a role-based matrix of who will be getting access to this new platform and what will be their access privileges.
	— Define and document a DR strategy (hot-hot, hot-cold, cold-cold, etc.).
	— Define and document how third-party tools will be authenticated and how they will access data within the platform (temp credentials, SSO etc.).
	— Define and document role-based, attribute-based, domain-based, tag-based data access, and sharing needs.
	— Define and document data consumption roles and policies, etc.

(*continued*)

Table 1-3. (*continued*)

Workstream	Main Responsibilities
Data engineering	— Design and document architecture for building a cloud-native and cloud-centric data lake strategy. — Design a framework for a reusable and repeatable data ingestion mechanism. — Design and document ingestion patterns and processes based on source types, source systems interactions, frequency (batch versus streaming etc.), data formats, and data types. — Design and document a framework for data cleansing, data quality assessment, and data validation and checks in an automated and reusable way. — Design and document a framework for data enrichment, data standardization, data augmentation, and data curation in a reusable and repeatable way. — Design and document a framework to capture the metadata of a business, operational, and technical nature and sync up with a catalog of choice. — Design and document a data reconciliation and audit balance framework for validating data loaded into the system. — Design and document a framework for building a data-reconciliation process for versioned datasets that might have changing dimensions. — Design and document a framework for building a business outcome (ETL) process in an automated and reusable way. — Define and coordinate with other teams to understand existing and "to be" engineering processes for DR strategy. — Define and coordinate with other teams to understand and engineer processes for the data access in an automated way. — Design and coordinate with third-party tools for data catalog, data governance, scheduling, monitoring, etc.

15

Define Common Terms and Jargon

Probably the single most important activity to kick off any project is the task that is needed to bring everyone on the same page. I have had challenges in my previous projects where we did not have a chance to align on the common terms and jargon. That always led to multiple issues and challenges for any technical discussion and architecture process throughout the project.

Here are a few examples where we wanted to align on this project:

- A common definition of data storage zones. Examples are *raw* versus *curated* versus *provisioned,* or *bronze* versus *silver* versus *gold.*

- Clear responsibility and features for the zones. Examples include what controls these zones should have versus what kind of data and life-cycle policies should the zones have.

- Common definitions for *tenant* versus *hub* versus *spoke.*

- Common definitions for *dev* versus *UAT* versus *prod* versus *sandbox* versus *playground.*

- *ETL* versus *ELT* with regard to the cloud platform.

- Common philosophy of loading data on-demand versus loading all data and processing on an ad hoc basis.

- Common philosophy for default personas and intended access control to data within the data lake.

This was an important alignment where we as a team not only interacted with customers for the first time, but we made great progress in terms of clearly documenting what was to be delivered in the subsequent weeks.

Key Takeaways

To recap, we met with all the key stakeholders including our sponsor for the data strategy work. We interviewed key personnel and identified key areas (to prioritize), and we understood the current landscape and maturity. We devised a strategy to work on three workstreams and defined key meetings and whiteboard sessions for the next few weeks (putting meetings on calendars for key personnel). Last but not least, we defined common terms and presented what our focus would be and the possible measure of success for this project.

Based on the series of discussions, in general our goal for the next steps were as follows:

> *Understand the customer's requirements*: The first step is to understand the customer's specific requirements and goals to develop a plan to achieve them. This includes understanding the data sources, data types, data volume, and other factors that may affect the design of the data lake.

> *Design the data lake architecture*: After understanding the customer's requirements, the next step is to design the data lake architecture. This includes selecting the appropriate storage technology, selecting the data ingestion and transformation tools, and designing the data flow and data management framework.

> *Develop the data lake*: Once the architecture is designed, the team can start to develop the data lake. This includes setting up the necessary infrastructure, building the data ingestion and processing pipelines, and managing the data lake.

Test and deploy the data lake: After the data lake is developed, it needs to be tested and deployed. This includes testing the data lake to ensure it meets the customer's requirements and deploying it in a production environment.

Monitor and optimize the data lake: Once the data lake is deployed, it's important to monitor its performance to ensure it's meeting the customer's goals.

CHAPTER 2

Enabling the Security Model

Objective: Identifying the Security Considerations

My responsibility as part of workstream was to define, design, and implement a holistic security model for the data platform.

My fundamental objective was to work closely with the customer's security and cloud engineering teams and with the AWS team to define a security blueprint that could help with the customer's platform, data, and application security considerations.

As we had already set up the important meetings ahead of time, we started having initial one-on-one meetings with each of the key security stakeholders (both internal and external) to document and design the key decision points (through knowledge discovery in data [KDD]) needed for designing the security blueprints. We eventually brought all the teams together to agree on the common solution and socialized the outcomes. This approach ensured we did not waste everyone's time and ensured we had targeted questions for specific groups and tangible outcomes designed and approved by each group.

© Nayanjyoti Paul 2023
N. Paul, *Practical Implementation of a Data Lake*,
https://doi.org/10.1007/978-1-4842-9735-3_2

The Recommendations

I used the following key design decisions to come up with a blueprint and ensured that those KDDs addressed the needs of each stakeholder. The objectives of the internal and external stakeholders were different. For example, the internal teams wanted a security blueprint that focused on a separation of concerns, the right access and security controls, and tight integration with enterprise security principles and policies, whereas the external stakeholders asked us to focus on cloud-native and best-of-breed technologies and the right service model to build the solution.

The following checklist was part of the vision alignment, and you can use this for your project too as a template. Be open to asking your own questions based on your customer's interest and their maturity (in other words, use this as a starting guide).

- PII columns: RBAC, ABAC features

- Central access control

- SAML vs. PING, etc.

- Strategy for data obfuscation

- GDPR and other data privacy

- Ownership of the platform, interaction with other stakeholders (CISO, legal teams, etc.)

- Legal/contractual obligations on getting/connecting data from a third party on the cloud

Each of these is detailed in the remainder of the chapter.

PII Columns: RBAC, ABAC Features

As we were bringing in data from third-party sources and external vendors, the chances of bringing in sensitive data is high. On top of that, the data is owned by different organizational units, which begs the question, is it OK for a group of people to have access to certain PII data that is specific to that organizational unit but cannot be accessed by other units?

The following are both the challenges and the requirements for PII column mapping from the requirements we received from the customer:

- Customers needed a single source of truth for all their data and analytical needs. Currently the data across the organization was siloed, which was one of the major reasons for this customer to venture into a data lake in the cloud. Hence, it was important for the data strategy to have open access to the datasets. However, the PII columns should be treated differently. The access to PII data should be based on a "need-to-know" basis.

- Each dataset needs to be tagged with a classification level, typically ranging from Confidential to Public. Confidential-tagged datasets have different encryption keys, and access to those datasets were on an on-demand basic (not open for all).

- Each column has a sensitivity level (typically L001, L002, etc.). These sensitivity levels should govern which columns of which datasets can be accessed by default versus which ones need special access.

- Datasets are organized as data domains (within business domains). Some of these datasets should be handled with the utmost care. For example, the HR team or finance team can access salary information, but other organizational/business units should not have access to it.

21

- Special roles and access grants should be allowed
  for accessing sensitive data. As this is an open data
  lake platform, personas such as data scientists or
  data analysts can request access to certain sensitive
  information based on business case and justification.
  Policies and processes should be in place to enable
  users and roles access to sensitive information for a
  specific duration of time.

- Access permissions should also be controlled based on
  the consumption and interaction pattern. For example,
  automated access to data for processing might have full
  access to all columns and datasets to ensure a quick
  and repeatable way of building data transformation
  and ETL jobs. However, the ad hoc and interactive
  access should be restricted based on the role and
  persona group the person/resource belongs to.

- Third-party tools that access data from the data lake
  should also respect the access control and guardrails
  defined in the data lake. They should impersonate
  the user who needs access to data via the third-party
  tools or have SSO integration for specific service role–
  based access.

Figure 2-1 provides a glimpse of the overall process that was followed
for this customer based on the AWS stack selected for the project. The idea
was to have a data strategy design (more to follow in the next chapters)
of organizing the structure of data into Raw (or Bronze), Curated (or
Silver), and Provisioned (or Gold) for the Automated (ETL jobs, etc.) and
Playground (ad hoc or interactive) access perspective. For the interactive
access process, the access control was defined at a granular level (tag,
sensitivity, and domain level) and was based on AWS Lake Formation

(based on the AWS technology stack selected). Access to any "curated" datasets had to be done based on automated policies registered while onboarding data in the data lake through the data pipelines. All these automated access policies were stored in the Lake Formation service of AWS centrally, and access to the data through any service (like Athena, Redshift Spectrum, etc.) was done through the Glue catalog managed and governed by the Lake Formation service.

let's

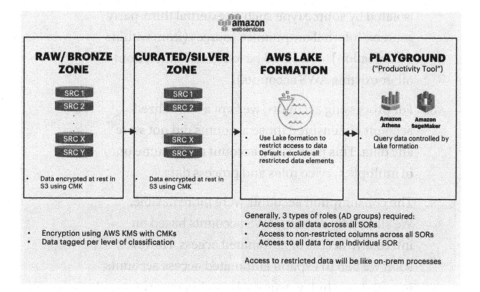

Figure 2-1. *A high-level view (with AWS technology stack) for a governed data lake*

We started the security and access control design by taking baby steps and handling the architecture on a use case by use case basis. We wanted to have a baseline architecture first and then test our hypothesis by laying out additional use cases and validating whether our architecture could stand the test of the same.

The following were the measures we took and the guidelines we followed to build the first iterations:

1. We created a multi-account AWS strategy to maintain the blast radius. The multi-account strategy can be based on three levels: ingestion accounts, processing accounts, and consumption accounts.

2. For the previous step, ingestion accounts were isolated by source type such as external third-party sources, internal on-premise sources (from within organization), and cloud based (data loaded from other existing AWS accounts).

3. For processing accounts, we kept a centralized account but ensured those accounts did not store any data. This processing account can assume one of multiple service roles and process data.

4. The consumption accounts were more flexible. We started by dividing AWS accounts based on interactive access or automated access. However, soon we had to expand automated access accounts into multiple hub versus spoke architecture as multiple organizational units wanted to own and manage their own "data products." Similarly, we had to scale up interactive access into multiple AWS accounts because of multiple independent teams and their needs to have a "self-service" capability for delivering business insights.

5. Once we decided on the AWS account setup, we tried to finalize the data encryption strategies. Each account had a bunch of AWS KMS CMK keys. We divided the keys into tier 1 to tier 3 keys. Based on the sensitivity of the datasets identified, we pushed data into independent buckets that had the default CMK keys associated with them. The service roles had access to those keys.

6. Once the encryption strategies were in place, we ventured into role-based, domain-based, and tag-based access control policies. Each dataset when being onboarded into the data lake was associated with three tags: business domain tags (like finance, marketing, general, etc.), data sensitivity tags (confidential, public, etc.), and column-level PII tags (L001, L002, etc., where L001 meant no PII, and L002 meant it has partial or entire PII information such as date or birth along with full name). We spent considerable time and effort discussing these with business and the CISO to come up with the tags.

7. Once the tags were in place, we introduced AWS Lake Formation. AWS Lake Formation is a service that allows a central access control and governance platform to enforce data access policies. Typically, Lake Formation ensures that the client applications (like AWS Athena, etc.) authenticates itself to access any data in S3. The authentication process grants temporary credentials based on the user's role. Internally, Lake Formation then returns only those datasets, columns, etc., that the user has "grants" for.

Hence, in our example, users who are from ROLE-A that belongs to ORGANIZATION UNIT (or business domain) B can query only those datasets that are tagged for ORGANIZATION B usage (or tagged for GENERAL usage). Additionally, the ROLE-A can view only those columns of the mentioned datasets that are tagged with either L001, L002, or L003 based on the tags allowed for ROLE-A.

8. Once the tag and role-based access were set up, we wrapped up the security and access control based on the consumption pattern. In this case, we focused only on AWS Redshift, and hence we defined policies for Redshift access and data sharing through IAM roles (more on Redshift in the upcoming chapters). Redshift was used to register data products that were owned by independent domain/business organizations, and we ensured the access control follows the same philosophy as mentioned earlier.

9. Lastly, we enabled the Playground area as a logical extension of the production setup. We enabled guardrails and processes to access data and services in the playground. This was mostly for data science interactive access. Chapter 5 talks about enabling the data science playground.

Figure 2-2 shows how the overall Lake Formation setup might look (from a high level).

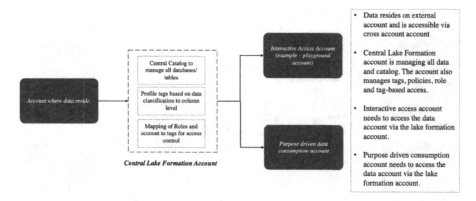

Figure 2-2. *How a central catalog and access control can be designed for managing role-based access for interactive users*

Central Access Control

Central access control is related (at least in this example) to the setup of the Lake Formation (centralized access control) AWS account, as depicted in Figure 2-3. Let's deep dive into what it means and why we designed it that way in this project.

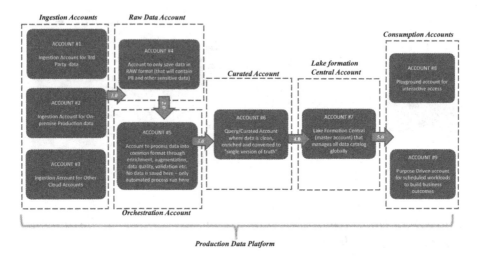

Figure 2-3. *A sample multi-account strategy for access control and separation of concerns to designing an enterprise-ready data lake*

Table 2-1 explains the choices made in this project.

Table 2-1. *Accounts Needed When Designing a Multi-account Enterprise Data Lake*

Account Type	Account Number	Account Purpose
Ingestion account	Account #1	• Only connect and access third-party data. • No access by any users to this account. Only ingestion jobs run in this account. No data is saved here. • Only ingestion-specific services are enabled.

(*continued*)

Table 2-1. (*continued*)

Account Type	Account Number	Account Purpose
	Account #2	• Only connect and access on-premise data. Data is saved with specific tiered encryption keys. • No access by any users to this account. Only ingestion jobs run in this account. No data is saved here. • Only ingestion-specific services are enabled.
	Account #3	• Only connect and access other cloud data. Data is saved with specific tiered encryption keys. • There is no access by any users to this account. Only ingestion jobs run in this account. No data is saved here. • Only ingestion-specific services are enabled.
Raw data account	Account #4	• Data is saved with specific tiered encryption keys. • No access by any users to this account. • No jobs run in this account; only cross-account access is provided for accounts #1, #2, and #3 to save data into this account and account #5 to read from this account.
Orchestration account	Account #5	• No data is saved into this account. • Only scheduled jobs run to clean up, enrich, augment, and validate data from account #4 and save to account #6.
Query/curated account	Account #6	• Data is saved with specific tiered encryption keys. • This account provides persona-based access to data.

(*continued*)

Table 2-1. (*continued*)

Account Type	Account Number	Account Purpose
Lake Formation account	Account #7	• Central security and audit account. • All data catalog and tables are registered here. • Policies for role-based, tag-based, and domain-based access are maintained here. • Central account to grant permissions as who can access which tables/columns/data based, etc. • Captures central audits.
Playground account	Account #8	• Enables interactive users to work with data. • Data scientists, data engineers, etc., have access to this account and they get cross-account access to account #6 based on the policies and permissions defined in account #7.
Purpose-driven account	Account #9	• Account where final consumption ready datasets reside. • All processes running here are scheduled and have a business reason. • No interactive or user-based access to this account.

Authentication and Authorization (SAML vs. PING, etc.)

This section is important for two reasons. Initially, documenting the Active Directory (AD) integration helps us map the users and roles to capabilities within the data lake as what the user can and cannot do. The other (and more important) part is the decision of who can see what data and how the user's role defines what domain/column-level data they can have access to. Table 2-2 lists what we discussed with our customer to understand their current approach and what kind of roles were needed for us to implement the access control process.

You can use Table 2-2 as a template and have similar documentation for your project scope for authentication and authorization policies.

Table 2-2. Technology Components, Services, and Roles Needed with the Corresponding Access Permissions

AWS Service	IAM	Type of Role	Permissions	Description
S3	External Process IAM User	IAM user	Write, create new subfolders.	Data stage to use the service account credentials to load data into landing zone.
	orchestrator-service-role1	Service role	Read from bucket.	Service role to access S3 landing zone.
	cloud-engineer	IAM role	Create bucket.	Cloud Engineering role will associate the ask to build, maintain, and manage cloud platform and infrastructure, etc.
	data-engineer	IAM role	Read from bucket.	Data Engineering/Cloud Engineering role will associate with Service role to read from landing.
	orchestrator-service-role1	Service role	Write, create new subfolders.	Read and write access to raw zone.
	cloud-engineer	IAM role	Create bucket.	Data Engineering/Cloud Engineering role will associate with Service role to read from landing.

data-engineer	IAM role	Read from raw zone.	Data Engineering role will associate Service role to read from landing zones.
functional analyst	IAM role	Read from raw zone.	Access raw zone for exploration.
data-scientists	IAM role	Read from raw zone.	Access raw zone for exploration.
orchestrator-service-role2	Service role	Read from raw zone.	Access raw zone for exploration.
data-analyst	IAM role	Read from raw zone.	Access raw zone for exploration.
functional analyst	IAM role	Read, write, create subfolders, share.	Owners of own home folders.
data-scientists	IAM role	Read, write, create subfolders, share.	Owners of own home folders.
orchestrator-service-role2	Service role	Read from raw zone.	Access raw zone for exploration.
data-analyst	IAM role	Read, write, create subfolders, share.	Owners of own home folders.

(continued)

33

Table 2-2. (continued)

AWS Service	IAM	Type of Role	Permissions	Description
	orchestrator-service-role1	Service role	Read to raw zone and write to quarantine zone.	Write to quarantine zone.
	cloud-engineer	IAM role	Create permission (for new resources in AWS platform).	Data Engineering/Cloud Engineering role will associate with Service role to read from landing.
	data-engineer	IAM role	Read from quarantine zone.	Data Engineering/Cloud Engineering role will associate with Service role to read from landing.
	External System IAM User	IAM User	Write, read.	Copy spec files to repo.
	orchestrator-service-role1	Service role	Read.	Read access to repo.
	cloud-engineer	IAM role	Create.	Data Engineering/Cloud Engineering role will associate with Service role to read from landing.
	data-engineer	IAM role	Read.	Data Engineering/Cloud Engineering role will associate with Service role to read from landing.

Step Functions	orchestrator-service-role1	Service role	Execute.	Execute step function.
	cloud-engineer	IAM role	Create, delete, monitor.	Data Engineering/Cloud Engineering role will associate with Service role to read from landing.
	data-engineer	IAM role	Create, delete, monitor.	Data Engineering/Cloud Engineering role will associate with Service role to read from landing.
Lambda Functions	orchestrator-service-role1	Service role	Execute, access CloudWatch.	Run Lambda based on new trigger file arrival in landing.
	cloud-engineer	IAM role	Create, delete, monitor.	Data Engineering/Cloud Engineering role will associate with Service role to read from landing.
	data-engineer	IAM role	Create, delete, monitor.	Data Engineering/Cloud Engineering role will associate with Service role to read from landing.
	orchestrator-service-role1	Service role	Execute, access CloudWatch.	Run Lambda based on new trigger file arrival in landing

(continued)

Table 2-2. (*continued*)

AWS Service	IAM	Type of Role	Permissions	Description
	data-engineer	IAM role	Create, delete, monitor.	Data Engineering/Cloud Engineering role will associate with Service role to read from landing.
	cloud-engineer	IAM role	Create, delete, monitor.	Data Engineering/Cloud Engineering role will associate with Service role to read from landing.
	orchestrator-service-role1	Service role	Execute, access CloudWatch.	Run Lambda based on new trigger file arrival in landing.
	data-engineer	IAM role	Create, delete, monitor.	Data Engineering/Cloud Engineering role will associate with Service role to read from landing.
	cloud-engineer	IAM role	Create, delete, monitor.	Data Engineering/Cloud Engineering role will associate with Service role to read from landing.
	orchestrator-service-role1	Service role	Execute, access CloudWatch.	Run Lambda based on new trigger file arrival in landing.
	data-engineer	IAM role	Create, delete, monitor.	Data Engineering/Cloud Engineering role will associate with Service role to read from landing.

cloud-engineer	IAM role	Create, delete, monitor.	Data Engineering/Cloud Engineering role will associate with Service role to read from landing.
orchestrator-service-role1	Service role	Execute, access CloudWatch.	run Lambda based on new trigger file arrival in landing.
data-engineer	IAM role	Create, delete, monitor.	Data Engineering/Cloud Engineering role will associate with Service role to read from landing.
cloud-engineer	IAM role	Create, delete, monitor.	Data Engineering/Cloud Engineering role will associate with Service role to read from landing.
orchestrator-service-role1	Service role	execute, access CloudWatch.	Run Lambda based on new trigger file arrival in landing.
data-engineer	IAM role	Create, delete, monitor.	Data Engineering/Cloud Engineering role will associate with Service role to read from landing.
cloud-engineer	IAM role	Create, delete, monitor.	Data Engineering/Cloud Engineering role will associate with Service role to read from landing.

(continued)

37

Table 2-2. (*continued*)

AWS Service	IAM	Type of Role	Permissions	Description
SNS	orchestrator-service-role1	Service role	Write.	Access SNS topic and write message with a predefined format.
	cloud-engineer	IAM role	Create, read.	Data Engineering/Cloud Engineering role will associate with Service role to read from landing.
	data-engineer	IAM role	Read.	Data Engineering/Cloud Engineering role will associate with Service role to read from landing.
Sagemaker	data-scientists	IAM role	Create, terminate, monitor.	Data Engineering/data scientists role will associate with Service role use this service.
	orchestrator-service-role2	Service role	Execute, access CloudWatch.	Execute Sagemaker notebooks for model implementation.
EMR (Spark)	orchestrator-service-role2	Service role	Execute, access CloudWatch.	Execute EMR jobs.
	data-engineer	IAM role	Create, terminate, monitor.	Data Engineering/data scientists role will associate with Service role use this service.

	data-scientists	IAM role	Create, terminate, monitor.	Data Engineering/data scientists role will associate with Service role use this service.
Athena	data-engineer	IAM role	Create tables, execute queries, access data from raw zone and other accessible S3 folders.	User's home schema for SQL-based exploration
	data-analyst	IAM role	Create tables, execute queries, access data from raw zone and other accessible S3 folders.	User's home schema for SQL-based exploration.
	functional analyst	IAM role	Create tables, execute queries, access data from raw zone and other accessible S3 folders.	User's home schema for SQL-based exploration.
	data-scientists	IAM role	Create tables, execute queries, access data from raw zone and other accessible S3 folders.	User's home schema for SQL-based exploration.
Redshift	data-engineer	IAM role	Create tables, execute queries, access data from raw zone and other accessible S3 folders.	User's home schema for SQL-based exploration.

(continued)

Table 2-2. (*continued*)

AWS Service	IAM	Type of Role	Permissions	Description
	data-analyst	IAM role	Create tables, execute queries, access data from raw zone and other accessible S3 folders.	User's home schema for SQL-based exploration.
	functional analyst	IAM role	Create tables, execute queries, access data from raw zone and other accessible S3 folders.	User's home schema for SQL-based exploration.
	data-scientists	IAM role	Create tables, execute queries, access data from raw zone and other accessible S3 folders.	User's home schema for SQL-based exploration.
DVL (EC2)	orchestrator-service-role2	Service role	EC2 boxes will have this role assigned.	Role associated with the EC2.
	data-scientists	IAM role	SSH, connect, run programs within EC2, access URL, download packages, access Athena, S3, Redshift.	Launch and run programs in EC2
	cloud-engineer	IAM role	Create 1 EC2 box with Tableau server, 1 EC2 for tableau desktop, 1 with Jupyter and other dependent packages.	3 boxes = 1 Jupyter, 1 Tableau server, 1 Tableau desktop.

	functional analyst	IAM role	SSH, connect, run programs within EC2, access URL, download packages, access Athena, S3, Redshift.	Launch and run programs in EC2.
	data-analyst	IAM role	SSH, connect, run programs within EC2, access URL, download packages, access Athena, S3, Redshift.	Launch and run programs in EC2.
End points needed: S3, Athena (JDBC), Redshift				
CloudWatch	orchestrator-service-role1	Service role	Write to CloudWatch.	Allow to write logs to CloudWatch group.
	orchestrator-service-role2	IAM role	Write to CloudWatch.	Allow to write logs to CloudWatch group.
	data-engineer	IAM role	Read CloudWatch logs.	Allow to write logs to CloudWatch group.
	data-scientists	IAM role	Read CloudWatch logs.	Allow to write logs to CloudWatch group.
	data-analyst	IAM role	Read CloudWatch logs.	Allow to write logs to CloudWatch group.

(continued)

41

Table 2-2. (*continued*)

AWS Service	IAM	Type of Role	Permissions	Description
	cloud-engineer	IAM role	Create log groups.	Create CloudWatch groups.
	functional analyst	IAM role	Read CloudWatch logs.	Allow to write logs to CloudWatch group.
Glue, Glue Crawler	orchestrator-service-role1	Service role	Execute glue and glue crawler jobs, run, execute.	Run glue jobs using this service role.
	cloud-engineer	IAM role	Create, terminate, monitor.	Create jobs, terminate, monitor.
	data-engineer	IAM role	Create, terminate, monitor.	Create jobs, terminate, monitor.
DynamoDB (Need VPC endpoint)	orchestrator-service-role1	Service role	Read, write to table.	Read and write spec file.
	data-engineer	IAM role	Read from table.	Read spec files for debug.
	cloud-engineer	IAM role	Create table.	Create table.
	orchestrator-service-role1	Service role	Read, write to table.	Read and write spec file.
	data-engineer	IAM role	Read from table.	Read spec files for debug.
	cloud-engineer	IAM role	Create table.	Create table.

* Table will have filename as partition key and version as sort key

SQS	orchestrator-service-role1	Service role	Publish and subscribe to queue.	Allow to publish message to the queue.
	data-engineer	IAM role	Read, list, view message in queue.	Allow to test and develop message and debug applications.
	cloud-engineer	IAM role	Create, monitor, manage.	allow for management.
	orchestrator-service-role1	Service role	Publish and subscribe to queue.	Allow to publish message to the queue.
	data-engineer	IAM role	Read, list, view message in queue.	Allow to test and develop message and debug applications.
	cloud-engineer	IAM role	Create, monitor, manage.	Allow for management.

Like with many identity providers (PING, OKTA, etc.), the goal is to have SSO set up with the AWS services so that users can log in to the AWS console using their existing credentials and AD roles. Once the credential and access setup are achieved, the final goal for the identity providers is to map the AD group and role to a corresponding IAM role within AWS so that the logged-in users have a well-defined access permission based on the role they belong to. In our case, the customer already invested in PING, and we worked on the previous matrix to map the user role from AD to the IAM role within AWS.

Strategy for Data Obfuscation

Data obfuscation is critical, especially when data is sensitive and belongs to financial or healthcare projects (like this one). In this project, we worked on multiple scenarios and approaches to data obfuscation based on timeline, customer expectation, and time to market.

Before we start, I want to differentiate quickly between *obfuscation* versus *encryption* versus *tokenization*, etc., as we spent quite some time with our customer using these terms.

There are multiple schools of thoughts on the differences and hierarchy between the approaches; however, the following is what we landed and agreed for the project.

- *Data obfuscation* is the process of making the data unreadable and unusable for normal processing. There are multiple ways of obfuscating the data.

  - *Encoding* is the process of translating the data to another representation of the same data. A simple way to achieve encoding is to change the character set from English to Spanish, etc. Encoding is one of the ways of making data obfuscated.

- *Masking* is the process of replacing the data with some "junk" values. Masking can be randomized or based on some mapping set and can be format preserving. Masking is another way of achieving data obfuscation.

- *Hashing* is the process of converting the value of data through a "statistical formula." Formulas like SHA1, HEX, etc., are examples where common data values are converted to a known outcome. The data from hashing is not human-readable but is easy to trace back. This is another way of data obfuscation.

- *Data tokenization* is the process of replacing the actual value of data with a "token." Typically, these tokens are unique for all the data being replaced, and the token to the actual data is ideally saved into some token vault. Because of the nature of tokens and the integrity it possesses, tokenized data can be used for analytics and can act as foreign/primary keys. Tokenized data is typically format preserving.

- Data encryption is the process of mathematically converting data (and possibly reconverting it to the original format) into something that cannot be used for any analysis. Data is typically encrypted with encryption keys and ideally stored in a secure location. Users (or roles) can have access to the encryption/decryption keys to convert and get the original data back.

Our journey, for this customer to enable data obfuscation, was based on legal obligations and contracts. A lot of data sources identified to be brought in were third-party sources, and those data providers had legal contracts with our customer to ensure no plain-text data would be made available in any cloud platform. Based on those guidelines, we wanted to follow the path of least resistance to ensure we could deliver the analytics platform (data lake) on time.

Customers already had an on-premise tokenization solution in place. Because of the restrictions, we could not have onboarded nontokenized data in AWS, which forced us to use the on-premise solution. Data was pushed into AWS (more on push versus pull later) through the data tokenization program, and the token vault was maintained on-premise. This ensured that only the "right" users who have an existing permission to the token vault can eventually see nontokenized data. This alleviated a couple of issues for us. First, we did not have to immediately solve the data tokenization problem in the cloud, and second, we didn't have to (re) solve the token access and permission issue.

In Chapter 4, I will discuss how we eventually moved away from the on-premise dependency on tokenization and introduced cloud-based access control based on tags, roles, domains, and data sensitivity.

A great lesson learned (and I follow for other projects as well) is to establish the customer expectations and agreed upon delivery time. It is OK to gather some technical debt provided we have business outcomes and other high-priority items to deliver.

GDPR and Other Data Privacy

We did not have to really worry about GDPR in this case as the customer was based in North America. However, we did talk about best practices and guidelines (and documented the same) to ensure that we could enable data privacy guidelines without having to redesign the whole analytics platform.

The fundamental policies for any data privacy act can be loosely associated with the following:

- Ability to manage and organize PII information within the data platform so that it is easy to take action on individual user information (like the right to erasure)

- Ability to encrypt data on individual level so that user data need not commingle unless approved/required

- Ability to organize data by groups (organization units, location etc.), so that user data can be classified and accessed in a restrictive way

- Have an independent encryption/decryption database or token vault to manage the keys/tokens and to manage user data across the platform so that the act of deleting the data from all systems within the organization can be as simple as deleting the keys/tokens

All these capabilities were addressed (as described earlier) by bringing in the right keys, using data obfuscation techniques, implementing RBAC and ABAC processes, and controlling the blast radius.

Having said that, as this project did not have to deal specifically with GDPR or other data privacy issues, we were content with documenting the findings and providing to our customer a list of guardrails and processes that could be established when needed to address any data privacy requirements in the future (see Figure 2-4).

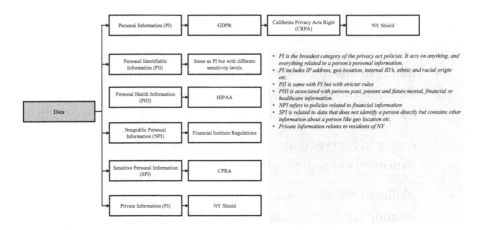

Figure 2-4. *Data privacy and regulatory compliance and governance needs*

Figure 2-4 is a quick reference to the kinds and types of PII regulations that we might need to consider for any project. This is not an exhaustive list, but we will use this information in the following chapters to help us connect the dots (based on types and patterns) and help take actions on specific PII regulations. For this customer, we focused on PII information only (however, for other projects we might need to use a combination of techniques).

Ownership of the Platform, Interaction with Other Stakeholders (CISO, Legal Teams, etc.)

As mentioned, although the project has an end goal and specific outcome to accomplish, it is important to investigate the big picture. It is critical that we know the overall enterprise goal and how other stakeholders are associated with that big picture.

As mentioned in Chapter 1, for this project there were many stakeholders with different measures of success criteria, and we wanted to understand that version from a platform ownership point of view.

This platform was sponsored by the business, which meant they were less critical of the technical implementation and more interested in the business outcomes. They were committed to building this "central data platform/hub," which would address the data and analytics issues across enterprises. We were planning to deliver and maintain the whole platform in phases. We were supposed to enable certain use cases for phase 1, but the overall goal was to have other organization units follow and be onboarded in the central data platform. As we were talking about onboarding multiple organizational units, getting approvals and aligning with teams like legal and CISO teams were critical.

Once we started aligning with the teams, we came across some business, security, and technical visions that shaped the key technical design decisions for our platform.

- The CISO team technical committee had heard about new updates and features that were supposed to be released soon (at that time) from AWS. Services such as Lake Formation with dynamic access control based on tags were important to them. They advised us to keep these features as part of our key design decision and put them in our backlog.

- As the enterprise strategy was to have a single data platform, it became clear that the setup needed to be distributed across multiple AWS accounts. So instead of us delivering the whole data platform on a single AWS account, we divided the whole solution into nine AWS accounts (more on this later).

- As we were supposed to onboard multiple teams and each team was supposed to "own" their own analytics, it became clear that the concept of a "data mesh" was important. Our design was very decentralized and democratized to enable this data mesh (again more on the data mesh design later).

- We already discussed the concepts of a "blast radius" and designated encryption keys for data at rest, etc. Those concepts were validated and audited (as they were critical and non-negotiable with the security and CISO teams).

Legal/Contractual Obligations on Getting/Connecting Data from a Third Party on the Cloud

This is something we already talked about, but I wanted to specifically call this out here to discuss the changing dynamics.

As mentioned, it was clear to us that there were third-party obligations not to have any unobfuscated PII data in the cloud from certain third-party data providers. The quick solution that we introduced was to use an on-premise tokenization process before onboarding data into AWS. However, this solution was only tactical. There were many reasons why this solution could not be a long-term solution for us. First, the objective for the customer was to shift the entire data and analytics platform into the cloud, which meant less dependency on managing the on-premise footprint. Second, the tokenization process was extremely slow and expensive. Any process that maps data and keeps a dedicated token vault will be a single point bottleneck for the whole data platform, and we wanted to mitigate that issue. Third, as we mentioned, cloud vendors like AWS were bringing in new technology such as RBAC, ABAC, and TBAC capabilities to enable fine-grain data management in the cloud.

For us, we knew what the future needed to look like (taking out tokenization and replacing that with a cloud-native solution). However, we needed to align with the legal team and have architecture whiteboarding sessions with third-party vendors to show the vision and inform them of our decisions. Of course, these discussions take time. So, our approach was to stick with the on-premise tokenization process for phase 1 release but then start having the discussion with the legal and architecture team within the customers and third-party vendors to address their concerns for our phase 2 plan. Luckily, we were able to achieve these as the project progressed, and we ended up enabling the cloud-native access processes (this solution is detailed in the following chapters).

This was a lesson for us, and I am sure all projects have these practical and unavoidable scenarios. The idea is to have both tactical and strategic solutions and guide the customers and vendors through that journey with the correct vision and partnership.

Key Takeaways

To recap, security is the single most important element for the data platform. Customers in general and the financial domain in particular need to be aligned to the best practices based on their organizational data protection and other data security policies and principles. In this journey of enabling the security, we had dedicated sessions with the customer's security groups including their CISO, security and cloud engineering teams, and external teams like the AWS services team.

Based on the series of discussions, we divided the solution blueprint into seven sections as described in this chapter. For each section, we detailed design sessions and documented the design and implementation methodology. Finally, once the security solution was ready, we focused on next steps, which were creating the organizational structure and roles, enabling the data lake, building out the playground, designing the DataOps and DevOps for production workloads, etc.

CHAPTER 3

Enabling the Organizational Structure

Objective: Identifying the Organizational Structure and Role

The organizational model helps us understand what we are committing to and where the solution will fit in. It also helps us understand the bigger picture and the delivery mechanism. It is important to understand and document the organization model and then lay it out in terms of work streams and deliverables.

The key objective of this section is to identify key roles, both from our side and from the customer's side, as well as their responsibilities, where they sit in the organization, and their ownership and escalation metrics. This chapter will also answer questions about the key roles needed to be staffed, their primary responsibilities, and their high-level job descriptions.

© Nayanjyoti Paul 2023
N. Paul, *Practical Implementation of a Data Lake*,
https://doi.org/10.1007/978-1-4842-9735-3_3

The Recommendations

Now, this comes as much from experience as science. The key roles, responsibilities, and positions along with the appropriate tasks and escalation metrics should be all captured here. Overall, once we have laid out this organizational structure, we know how the teams will be organized, what they need to "own," who the leads are for that team, and who the leadership is from both the project and the customer's point of view. Most important, it gives a structure to the whole project from a resources and people point of view.

For this project, we have the setup shown in Figure 3-1.

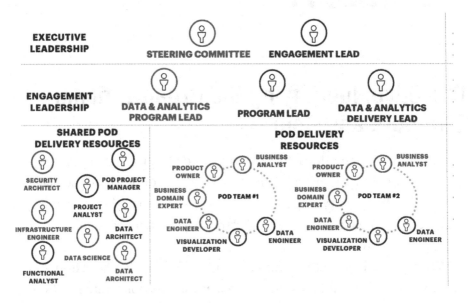

Figure 3-1. *Example team structure for project delivery*

Example Template for the Project

The key takeaways from Figure 3-1 are the roles and key ownership in the delivery model.

This is an example; however, the point is to discuss with customers what areas they want to bring their expertise to and how we can augment/support them. In this project, the customer wanted to own the platform and key design decisions along with the security and infrastructure.

Tables 3-1 and 3-2 provide a guide to the roles we filled in and the expectations of the roles from the customer. This separation of concerns keeps the workstream setup and task allocation and dependency management organized.

Table 3-1. *Key Roles Needed from the Consulting Company Side (Our Side)*

Key Role (from Our Side)	Role Description
Engagement lead	Senior leadership and executive escalation
Program lead	Manage issues and escalations and project status
Data and analytics delivery lead	Accountable for overall delivery of the engagement
Pod project manager	Manage day-to-day delivery of the work activities
Data and technical architect	Define and design data architecture and technical architecture including data science architecture and processes
Functional analysts	Manage use case–related business and functional requirements
Project analysts	Overall project analyst support
Visualization developers	Develop visualization engineering
Data engineers	Source and curate data, develop data marts
Data governance consultants	Overall data governance support and guidance
Data scientists	Develop machine learning use cases and models

Table 3-2. *Table Describing Key Roles Needed from the*
Customer Side

Role (from the Customer Side)	Role Description
Data and analytics delivery lead	Accountable for overall delivery of the engagement
Data architect	Define and approve data architecture
Product owner	Own the project scope and outcome
Business analyst	Own the project requirements elicitation
Business domain expert	Provide domain expertise on use cases in scope
Infrastructure engineer	Guide cloud infrastructure, including code promotion between environments
Data science lead	Provide overall guidance on data science use cases and methodology
Security engineer	Review and approve data classifications and storage on the cloud

Additionally, the organizational model was important to help us understand where each stakeholder sits within the customer organization and who to reach out to during the project for approvals and key decisions.

Another reason it was important to ensure we had the right organizational alignment was the fact that our customer had existing business processes already running on-premise. For our customer to move to the cloud, we had to ensure that those business processes (reports, data marts, machine learning models, etc.) had the same outcome in the cloud as they did on-premise so that the business could have the same "trust" on the outcomes.

A big part of that trust was to correctly document the business process, understand the current business logic, and set up a baseline to measure against. For our customer, the on-premise solution was based on a data warehouse on the Oracle and DB2 systems. We aligned with the developers and leads from the enterprise data warehouse (EDW) team. The development team was set up jointly with key resources from the EDW side (Oracle and DB2 engineers) for this cloud modernization journey. Our business analyst and developers sat with the EDW team to profile the jobs and detail the technical and business requirements. Those requirements were then mapped to the business logic and then the technical implementation guidelines and test cases. The outcome from that exercise ensured that all the key business processes were mapped and documented with the proper approval and sign-off.

Additionally, Table 3-3 is a quick operational model from the functional area. This table provides a quick guide to the key roles (from our consulting side) and the business functions that they belong to. The key responsibilities also map to the key objectives and actions that the functional roles will own.

Table 3-3. *Key Roles and Their Responsibilities*

Key Functional Roles (from Our Side)	Business Function	Key Responsibilities
Data engineer	Technology/ exploration	Moving data from on-premise to the cloud Limited assistance in staging/enriching data Support data steward in maintaining exploratory data catalog Testing and creating transformations to be applied to landing/testing and creating transformations to be applied to exploratory zone data in order for it to be promoted to the provision zone Support data scientist by creating datasets for discovery and model building
Data owner	Data management	Providing authorization for usage of data for specific analytics use cases Responsible for data quality of their assets Responsible for identification and protection of sensitive data for their data assets Monitoring the usage and access of their data assets (by reviewing access reports provided by the platform owner)

(*continued*)

Table 3-3. (*continued*)

Key Functional Roles (from Our Side)	Business Function	Key Responsibilities
Data steward	Data management/ exploration	Approving usage of data for specific analytics use cases based on policies set by the data owner Providing business metadata (classification, business terms, sensitive data info, etc.) for their datasets Defining business rules for DQ assessment for their datasets Cataloging exploratory data and lineage
Data scientist	Data management/ exploration	Hypothesis and experiment design Discovery, analyses, model building Visualization design Understanding data in exploratory and provision zones
Data analyst	Data management/ exploration	Discovery, analyses, model building Visualization and dashboard build
Business analysts	Exploration	Ad hoc reports

(*continued*)

Table 3-3. (*continued*)

Key Functional Roles (from Our Side)	Business Function	Key Responsibilities
Security admin	IT/network/ security	Manage the overall firewall and access control into the environment Define security policies for data encryption, AD server Define cloud infrastructure design (subnets, VPCs)
Platform admins	IT	Administrate the overall environment, services, and access control Define environment usage policies and security Create and update IAM users' access
Platform operations	IT	Monitor the overall environment and usage Download the S3 access logs into Splunk Publish data access reports

Once the organizational model is established, the key outcome of the process should be to map the operating model and responsibilities into a low-level delivery plan that will be jointly owned and accepted by the customer and service provider team (our team). At a high level, we had the following plan (this can be used as a template for other projects):

- *Define the scope and call out success criteria*: To be jointly owned by the customer and service provider team (my team)

- *Define high-level cloud architecture*: To be owned by the service provider

- *Finalize the architecture and sign-off:* To be owned by the customer

- *Finalize the technology stack and suggest POC if needed:* To be owned by the service provider team

- *Accept and sign-off on POC and technology stack:* To be owned by the customer

- *Enable cloud services (in dev, UAT, prod):* To be owned by the customer team (with support provided by the service provider team)

- *Setup of the platform access and roles (including AD integration, etc.):* To be owned by the customer team (based on the security model shared by the service provider team)

- *Design for data management capabilities such as ingestion, standardization, data quality, data reconciliation, data conformation, building data products, etc.:* To be owned by the service provider team

- *Building unit and integration test suites:* To be owned by the service provider team

- *Security design and architecture:* To be owned by the service provider team (with support and approval from the customer)

- *Deployment of the security framework and enablement of key security services and measures:* To be owned by the customer team

- *Data source profiling and documenting key business processes:* To be owned by the service provider and customer team jointly

- *Delivering value through data analytics and insights*: To be owned by the service provider team

- *Overall solution acceptance and sign-off*: To be owned by the customer team

Key Takeaways

To recap, we cannot start the solution design and build unless we have a proper responsibility, accountability, consulted, and informed (RACI) matrix. The matrix provides the key roles with their description and their focus areas that we need to fill. This also provides details from the customer's side about the key positions to provide guidance, thought leadership, and sign-offs.

Once we have this, we can focus on the next part, which is the data lake design and implementation.

CHAPTER 4

The Data Lake Setup

Objective: Detailed Design of the Data Lake

This is the most technical part of the project. This is where we design and
deliver the working solution that provides the business value. Once we
have established the key processes, personas, roles, and responsibilities
and have divided the areas of work into a proper cadence, this is the phase
where we start building things and delivering value to our customers. This
is the part where things start to take shape.

As I started this phase of my journey, I had a clear line of sight. By
now, I was able to build a good relationship with the stakeholders, and the
process I followed ensured that I cover all the bases. Again, these are my
recommendations based on my experience. When you embark on your
own journey, keep the overall objective in mind. We as practitioners have
the tendency not to see the bigger picture.

The Recommendations

In this chapter, I will initially focus on the key activities that I enabled for
the project with respect to structuring the data lake. Here, the focus will be
on the data lake implementation and the surrounding data management
principles.

© Nayanjyoti Paul 2023
N. Paul, *Practical Implementation of a Data Lake*,
https://doi.org/10.1007/978-1-4842-9735-3_4

The following are the recommended key design considerations and decisions for building a data lake from the ground up:

- We will structure the different zones in the data lake.

- We will define the folder structures and a hierarchy for the zones.

- We will manage data sensitivity as part of the folder structure design.

- We will discuss encryption/data management keys for organizing data.

- We will discuss the overall data management principles.

- We will discuss the data flow.

- We will set the access for each zone.

- We will discuss the file formats and structures in each zone.

Structuring the Different Zones in the Data Lake

The data lake should be a central, global, and trusted repository of data. The data that is loaded into the repository is enhanced and augmented to become reliable and trustworthy. This process of converting the data from the "system of records" into a reliable set of information is done in multiple hops. Different schools of thought have different recommendations as to how many hops data lakes should have. My take is to start from the fundamentals and add more hops as needed for a business context (as no two data lakes are ever the same).

Companies like Databricks and others have discussed the concepts of lake houses and data lakes in blogs and whitepapers. That documentation provides the right baseline for organizing the data. At a minimum, data lakes should be organized into four hops: Raw (or Bronze), Curated (or Silver), and Provisioned (or Gold), along with an optional Playground (or Exploratory). The ideology is quite simple. The Raw zone (or hop) provides a location for data to be housed as is. The data is stored in its original format to avoid risk of reloading data from external systems. The Silver layer is where data is cleansed, standardized, enriched, validated, audited, and augmented. This converts data from the Raw bucket into a "single source of truth." The data is typically saved in a common format. The Provisioned zone is purpose- and use case–driven; datasets lying in this zone are typically associated with a subject area or use case and typically go through an extract, transform, load (ETL) process with the appropriate business logic. The big difference between Curated and Provisioned is that Curated datasets are never joined against each other to solve any business problem, whereas the Provisioned datasets are always a representation of joins between multiple Curated datasets (based on a business use case). Another key difference is that typically the datasets in Raw and Curated have a one-to-one mapping, whereas the datasets between Curated and Provisioned have a many-to-one mapping.

Additionally, in most projects, we enable a special zone called Playground (or Exploratory) to provide role-based and persona-based interactive access capability on Curated and Provisioned data to ensure users can explore, be creative, and come up with next best model or business insights through data analysis. There are many differences between the Playground and Provisioned zones; the primary ones are specially related to interactive versus automated access and shadow IT. Typically jobs running on Provisioned are scheduled and noninteractive and managed by IT teams, whereas the Playground zone is for interactive purposes, and no work done here is scheduled or maintained by IT. Figure 4-1 is a quick recap of these points.

Figure 4-1. *Overview of dos and don'ts in each section/layer of the data lake*

The previous points describe the organization of data in zones; these zones provide some guiding principles, but we need more structure to organize data and propagate the data across zones. In the next sections, we will deep dive into the processes shown in Figure 4-2.

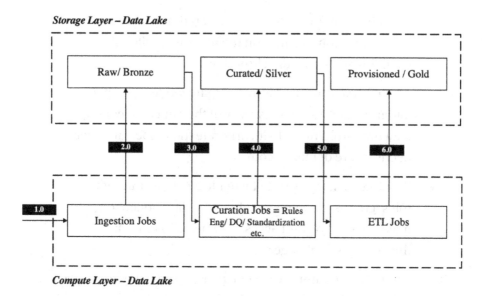

Storage Layer – Data Lake

Compute Layer – Data Lake

Figure 4-2. *Simplified view of the data lake processes and how each zone/layer is populated from the previous one*

Defining the Folder Structure of the Zones with a Hierarchy

Once we have defined the zones, the next step for us in the project was to decide on the principles of organizing data in these zones based on the source systems. Typically, there are many ways to organize data into the previously mentioned zonal structures. The following are some of the principles that have worked well for me in my past projects. The reasons why you should spend some time to think through the folder structures and organizations of the data are as follows:

- The data needs to be controlled per the source so that data from each source can be controlled differently (through separate encryption keys, default behavior, etc.).

- It is important to control the "blast radius." This will ensure that compromise on the data store will not result in large-scale data breach issues.

- A lot of times data needs to be organized using a special folder to capture the sensitivity of the dataset. Keeping data organized helps to incorporate those special folder needs (more details to come).

- Eventually, data needs to be made available through roles, groups, domains, etc. Keeping data organized makes it easy to grant access and helps to minimize those access challenges.

Based on these conditions, we came up with the following suggestions for the customer. As the data was coming from multiple sources (both structured relational sources and nonrelational sources like APIs), we enabled the following strategy to organize the data.

Structuring Data from Relational Stores (Raw Zone)

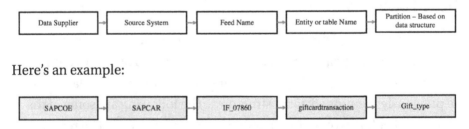

Here's an example:

This example shows how we proposed to structure data coming from relational systems. The idea is to divide the Raw zone into five levels. The first is the data supplier name (example, SAPCOE points to the SAP system), the second level is the source system name (like SAP CAR), the third level is the feed name itself (like IF_07860), the fourth level is the actual entity name or table name (for example, the gift card transaction table), and finally the last folder structure is either a default YYYYMMDD (or YYYYMMDDHHMM) partition folder based on the load date or some other structure that is specific to one or many column values of the incoming data (such as the customer state, ZIP code, etc.). This last layer ensures that in case of issues when loading data for a particular day (or hour), the data can be deleted and reloaded without impacting existing data already in the systems.

Structuring Data from Relational Stores (Curated Zone)

Here's an example:

The idea for the Curated zone structure for the relational data is to divide the Curated zone into five levels. The first is the data supplier name (for example, SAPCOE points to the SAP system), the second level is the source system name (like SAP CAR), the third level is the feed name (like IF_07860), the fourth level is like the actual entity name or table name (for example, the gift card transaction table), and finally the last folder structure is either a default YYYYMMDD (or YYYYMMDDHHMM) partition folder based on the load date or some other structure that is specific to one or many column values of the incoming data (such as the customer state, ZIP code, etc.). This last layer ensures that in case of issues when loading data for a particular day (or hour), the data can be deleted and reloaded without impacting the existing data already in the systems.

Structuring Data from Relational Stores (Provisioned/Gold Zone)

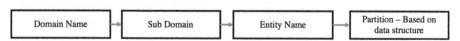

For the Provisioned zone, the idea is to organize data by domain or subject area for a specific use case (or business process). Not all data that is in the Curated zone will be available in the Provisioned zone. If there is a use case that needs to be solved, then the use case will be represented in the Provisioned zone. Typically, the use cases are "owned" and "managed" by a specific domain (like marketing/finance, etc.); hence, the structure of the data in the Provisioned zone needs to be representative of the ownership and management of that use case.

For example, if we plan to build a customer 360-degree data model that takes data from the Curated/Silver zone, then we can implement an ETL script that loads the four tables from Curated and saves the result into the Provisioned/Gold zone table like Marketing_Domain/Customer_Domain/Customer_360_Table/partitioned by Customer ID for quick search.

Here is the structured data from the external APIs (nonrelational for Raw and Curated):

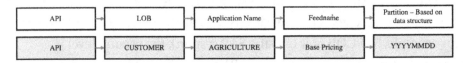

In theory, the first level can be the source name itself (like the actual API name or other external data source structure for Raw and Curated remains the same), the second layer can be the line of business (LOB) like CUSTOMER here), the third layer can be the application name within the API (like API that provides agriculture equipment pricing versus nonagricultural equipment pricing), the next layer can be the feed name (like the base pricing API), and finally the last one is the partition folder structure like YYYYMMDD (as discussed earlier).

The Provisioned zone for the API remains the same as the relational datasets as they will still power some business use cases.

Another view of the folder structure can be to organize data as a warehouse design, like many customers use an extended version of a lake house to organize data in Raw, Curated, and Provision in a warehouse design (although I don't think this is as flexible as the top one) where we are organizing by LOB-ApplicationName-FeedName-Partition, as shown in Figure 4-3.

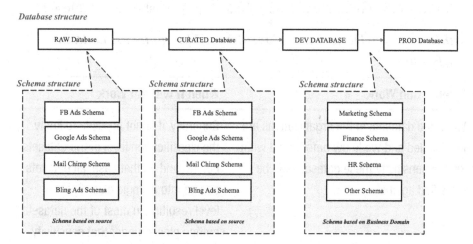

Figure 4-3. *How the Gold zone is organized across the Raw and Curated zones in the development (dev) and production (prod) environments*

If we look closely, the concepts are the same. Only the representation changes.

Managing Data Sensitivity as Part of the Folder Structure Design

This is one of those edge cases. We briefly touched upon the concept of introducing a static sensitivity of data in the folder structure itself. Let's see how this will look and some advantages and disadvantages of it.

```
DOMAIN_NAME / SUB_DOMIAN_NAME / <S1 or S2 or S3 or S4 as
sensitivity> / ENTITY_NAME / PARTITION_STRATEGY
```

The previous is a depiction of how we can introduce a special folder above the entity (or table name) to represent if the table is sensitive or not. Typically, this folder can have values like S1/S2/S3/S4, which can represent the typical sensitivity paradigm like PUBLIC/CONFIDENTIAL/RESTRICTED/PRIVATE. Some of the reasons why this kind of static sensitivity might or might not work for a project are shown in Table 4-1.

Table 4-1. *Key Decisions to Consider for Managing PII Data in Each Zone*

When It Can Work	When It will Not Work
When the datasets in the organizations are pre-labeled and we know beforehand what kind of sensitivity these datasets can be classified against.	Generally, it is not possible to know the classification level of the datasets beforehand. In that case, all datasets can go into a single classification level resulting in most of the datasets getting placed under that sensitivity.
When datasets have an equal distribution of classification so that there are enough datasets in S1 to S3 levels. If the datasets are all skewed in one specific classification level, then users who do not have access to that classification level cannot perform any substantial work as they will have no access to most of the data.	Continuing from the left side, in the current project, 95 percent of the dataset (as the project was financial in nature) has one or more columns marked as PII, which resulted in 95 percent of the datasets classified as S4. If we introduced the sensitivity static folder, 95 percent of the datasets would have ended up in the S4 classified folder thereby making this extra folder useless (as remaining 5 percent of data cannot practically be used by someone without having access to the other 95 percent of data).

(continued)

Table 4-1. (*continued*)

When It Can Work	When It will Not Work
With the introduction of some of the security services like Lake Formation, etc., it became easier to apply dynamic policies based on dataset tags rather than having static predefined folder design.	

Because of these reasons, for this project, we decided to go without adding an additional static sensitive layer in the folder structures.

Setting the Encryption/Data Management Keys for Organizing Data

In this section, we will discuss two specific things. One is the process of managing the "blast radius" based on transparent and at-rest encryption, and the second is to manage custom encryption. We talked about the obfuscation process and general encryption (and the variations) in Chapter 3, so in this section we will talk about transparent encryption/decryption.

AWS Key Management System (KMS) allows you to manage cryptographic keys to manage access to data and other services. KMS can be used to encrypt/decrypt data within a data lake and can also be used to generate other keys (data keys) that can help to encrypt external data and also implement custom encryption processes. There are three kinds of KMS setup that can be enabled, as follows:

Customer Managed Keys (CMK): Keys that customers create on their own but allow AWS to manage. Customers have full access to those keys in terms of maintaining them, disabling them, rotating them, granting permissions on them, etc. For this customer implementation, we used CMK as we wanted to have that tight control.

AWS Managed Keys (AMK): Keys that are created, managed, rotated, and used by AWS but for a specific account. Customers do not need to worry about the key management, key rotation, etc. It is an easier option than to bring our own keys.

AWS Owned Keys (AOK): Keys that are managed centrally by AWS and can be used by AWS across multiple accounts. This can be typically used for development accounts just to lower the cost and management/maintenance of keys.

Additionally, AWS KMS provides these three kinds of keys as symmetric keys (when we use same keys for encryption and decryption), asymmetric keys (which has public and private keys), and data keys (which can be used to custom-encrypt a large number of datasets), which provides a downloadable key to be used independently. See Figure 4-4.

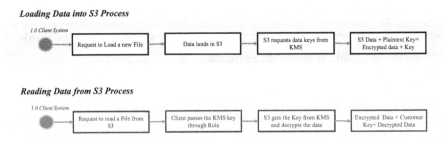

Figure 4-4. *Simple encryption and decryption process of AWS KMS*

Based on these pointers, we went with the CMK for this project implementation. We already talked about the folder structure with the five levels of hierarchy as DOMAIN_NAME/SUB_DOMIAN_NAME/ENTITY_NAME/PARTITION_STRATEGY. To take advantage of managing the "blast radius," we ensured that we have one CMK per DOMAIN as follows:

> DOMAIN_NAME1: CMK1
>
> DOMAIN_NAME2: CMK2
>
> DOMAIN_NAME3: CMK3, etc.

In this structure, we tagged the CMKs with the roles and the IAM policies for AWS to ensure only the right personas have the right access to certain DOMAIN datasets. Again, we could have made one CMK per SUB_DOMAIN_NAME (or ENTITY_NAME), but we were mindful of the fact that CMKs come at a cost, and also the CISO was happy enough (for this customer) about the separation of encryption per DOMAIN level.

Quick FAQs on the Data-at-Rest and Data-in-Transit Encryption

Table 4-2 describes data-at-rest encryption.

Table 4-2. *FAQ on Data at Rest*

What is it?	Data encryption at rest allows data in buckets such as Landing, Raw, and other zones to be encrypted transparently.
What does it cover?	Data-at-rest encryption typically covers encryption of data within buckets, databases, data stores, and data warehouses.
What does it solve?	Data encryption at rest provides two types of security process. <ul><li>It protects against disk theft or external hacks on the physical hardware.</li><li>It protects by ensuring the right roles can have permission to see the actual data.</li></ul>
How can it be implemented?	Most cloud vendors have three ways of enabling data-at-rest encryption. <ul><li>Default cloud-based encryption like AES</li><li>Cloud-based account-specific keys that can be generated and used to encrypt data</li><li>Bring your own CMK and use that to encrypt data in buckets</li></ul>
What are the data encryption models?	Data encryption can be achieved using either client-side or server-side modes. <ul><li>Server-side encryption is the simplest and most transparent way. The underlying data storage layer is responsible for encrypting the data.</li><li>This is the most common and most usable way to perform data encryption at rest.</li><li>Client-side encryption is managed by each application. The applications are responsible for encrypting/decrypting data when storing or retrieving it.</li></ul> This process has overhead to manage the encryption process independently for each application that interacts with the data.

Table 4-3 provides information about data-in-transit encryption.

Table 4-3. *FAQ on Data-in-Transit Encryption*

What is it?	Data-in-transit encryption is a way to ensure that the data is encrypted while being transferred over network between applications to data storage.
What does it cover?	It covers all data transfers over internal or external networks.
How can it be implemented?	• Most client vendors provide default TLS/SSL encryption for data movement within the network. • HTTPS endpoint for all data storage access. • Encrypted SSH connection to any VM to the cloud. • VPN gateway to protect network connections between on-premise and on-cloud interactions.

Looking at Data Management Principles

We talked about the concepts of zones, we discussed the folder structure and strategy, and we even focused on the concepts of managing the sensitivity and encryption mechanisms. Now, let's talk about what needs to be productionized and how we can solve our customer's business problem. This is where we start talking about building solutions and deploying with a production-ready data pipeline.

Let's start with the question of what should be developed once the platform, security, infrastructure, and people have been identified. We want a robust process that is repeatable and reusable so that we can start churning out really good data engineering pipelines that can help bring in the data, enrich the data, validate the data, and wrangle the data to produce the goods that drive business decisions.

From my experience, we should think of building the data engineering solution in three ways.

First, focus on building a robust foundational structure for the data pipelines. Second, focus on building business processes that can be developed on the foundational structure that is repeatable (for building data and ML products) and that can drive business outcomes. Finally, the enable data democratization so people can be "creative" and make the organization really "data driven." From my experience, the overview shown in Figure 4-5 provides the basic building blocks of what the data management process should consist of.

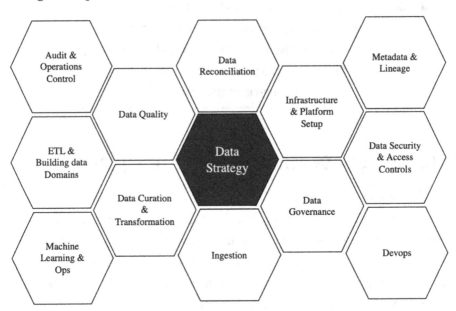

Figure 4-5. Overview of all modules/components needed to be addressed to implement the data lake

Table 4-4. *Modules*

Data Strategy Component	Key Decision Points to Consider
Infrastructure and platform	Decision on using SaaS versus PaaS Decision on serverless Decision on custom versus products Decision on structure of data lake (number of zones, data progression, etc.) Decision on data classification process and how it impacts the structure of data in the lake (in terms of managing encryption keys, etc.)
Data governance	Who are the primary stakeholders Decision around definitions of certain roles in the data lake (data analysis, engineers, business, etc.) Define process around onboarding new data, exposing new data products, and onboarding different roles in the data lake Define process around how business will interact with data and data services
Ingestion	Decision around pull/push Decision on common data formats Decision on acceptable SLA on bringing data Decision on failover, auto scalability, and workload management based on prioritization, time, and use case Decision on framework-driven approach to standardize ingestion processes

(*continued*)

Table 4-4. (*continued*)

Data Strategy Component	Key Decision Points to Consider
Data curation and transformation	Decision around rules engine, compliance, enrichment of data
	Decision around how and what triggers the rules engine
	Decision around outcomes and decisions by the rules engine
	Ease of use and ability for business to use as self-service process
Data quality	Decision on common framework and policies to apply for data checks and validations
	Decision to act on data quality validations (quarantine, etc.)
Data reconciliation	Ability to perform audit balance checks across sources and targets
	Ability to handle multiple reconciliation KPIs and metrics
Metadata and lineage	Decision on what and how to capture all metadata
	Decision to keep track of data lineage
	Decision to create centralized data catalog
	Decision to expose metadata to data lake users for search, discoverability, etc.
Data security and access controls	How many roles?
	Who gets to access which zone within the data lake
	Who can see what classification level of data
	How to manage user onboarding etc.
DevOps	Automation around productionizing data pipelines
	Scheduling

(*continued*)

Table 4-4. (*continued*)

Data Strategy Component	Key Decision Points to Consider
Audit and operations control	Keep track of jobs running, success, failure
	Manage information and handling of job status, rerun capability, etc.
	Integration with a centralized audit management system (like NOC, etc.)
ETL and building data products	Choice of technology
	Reusability and repeatability
	Decision of data models
	Ability to connect and populate the data models
Machine learning and ops	Define experimentation zone
	Define experiment and trials
	Define data access based on roles and data classification levels
	Move from experimentation phase to productionizing phase

Let's talk about each of the components of data management in the next section, which explains the overall data flow and how the data pipelines need to be designed and managed.

Understanding Data Flows

This section will define how we eventually designed and implemented the data plumbing based on the components described earlier. Before explaining each of the components, let's look at the big picture.

Figure 4-6 explains how the components can be stitched together to form a pipeline that can process data and create data or machine learning products that provide value to the business. Again, this is a representation (of the client implementation) and can be leveraged as a reference for other projects.

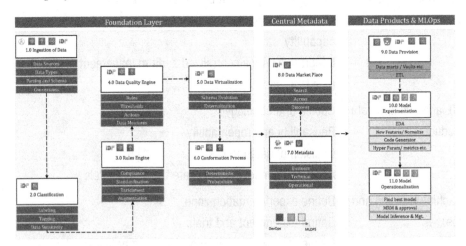

Figure 4-6. *End-to-end data flow for the data lake*

We implemented the previous solution using AWS native capabilities for this customer as the alignment was to be as native and as serverless as possible.

The data flow was to enable ingestion capabilities (through multiple source connectors, data type handling, schema mappings, and conversions) by bringing in data from external and internal sources into the data lake. Once done, we executed a classification process to start labeling, tagging, and identifying the sensitivity of the datasets. The classification process was custom implemented using pattern matching, known schemas, and some NLP libraries. Next, we executed the rules engine to fix data issues through standardization, lookups, enrichments, augmentation with industry standards, etc., to bring data into a more usable format. After the rules engine, we executed the data quality engine to validate and check the dataset and ensured that there were no anomalies that could impact the downstream applications. The DQ engine can take action around moving impacted records into a special zone (quarantine) so that those data issues can be fixed later. Next, we enabled the data virtualization module, which can be as simple as creating a catalog and enabling SQL-like access to standardize data for more democratized access.

After the virtualization process, we enabled the confirmation process that can consolidate and merge datasets based on the changing dimensions of the data. We enabled slowly changing dimensions (SCDs) 1 and 2 to confirm the data for downstream applications. Once the data was confirmed and virtualized, we enabled a metadata sync-up and pushed all the metadata information into a central catalog that provided a "data marketplace" experience for the customer. With a data marketplace experience, the "right" personas can start using the data lake to understand which datasets reside where; how they are organized; what labels, tags, sensitivity, etc., are associated with each dataset; and what rules (standardization and DQ) were used to fix the data.

Next, we enabled the ETL capability for building data and ML products through easy and intuitive data wrangling and templatized jobs. Lastly, we enabled the MLOps process for the customer (not scoped for this book and surely in my bucket list for the next one) to build and deploy models at scale to solve a business problem that is more predictive in nature.

However, it is always important to know the tech stack operating in the same plane. Table 4-5 is a quick description of the tools we discussed with the customer before finalizing the AWS native stack.

Table 4-5. *Key AWS Services Needed for Implementing Each Component of the Data Lake*

Requirement	Key Features	AWS/Partner Tools	Third-Party Example Tools
Ingestion	• Ability to handle all kinds of data • Ability to handle changing schema/evolution of schema • Ability to handle structured/unstructured data • Ability to handle delta changes • Ability to integrate with metadata management systems • Ability to integrate with data governance solutions • Ability to be extended, customized easily • Ability to handle complex data structures and schemas • Ability to be automated • Integration with error management, log management • Ability to version data and code • Ability to optimize and manage formats like Parquet, Avro, orc, etc. • Ability to be executed as config-driven reusable asset • Ability to be run as serverless module if needed	• AWS Glue • AWS EMR • Step Functions • Lambda • API Gateway • Firehose • Kinesis Streams	• Talend • Informatica • Streamsets • Nifi • Kafka

(continued)

Table 4-5. (*continued*)

Requirement	Key Features	AWS/Partner Tools	Third-Party Example Tools
Metadata Management	• Ability to capture business, operational, technical metadata • Ability to profile and catalog the data • Ability to be searchable and discoverable • Ability to extend as API • Ability to capture metadata across data lifecycle • Ability to capture metadata across models and data science • Ability to identify sensitive data and label automatically • Ability to document model experiments and processes • Ability to draw lineages and data progression steps	• Glue Catalog • Elasticsearch/ DynamoDB	• Alation • Collibra • ASG

(continued)

Table 4-5. (*continued*)

Requirement	Key Features	AWS/Partner Tools	Third-Party Example Tools
Data Curation	• Ability to handle complex data transformations • Ability to express and execute transformations as rules • Ability to integrate with metadata for policy-based rules • Ability to handle compliance, standardization, normalization rules • Ability to integrate with metadata • Ability to build and manage data marketplace • Ability to be config driven and reusable • Ability to work as part of automated pipelines and scheduled	• AWS Glue/EMR • Step functions	• Talend
Data Conformation	• Ability to linkage data • Ability to perform entity resolution • Ability to perform harmonization of complex data sources • Ability to generate holistic data profiles	• AWS Glue/EMR	• IER • Tamr • Paxata • Talend

(*continued*)

Table 4-5. (*continued*)

Requirement	Key Features	AWS/Partner Tools	Third-Party Example Tools
Data Quality	• Ability to identify quality gaps in data • Ability to identify and report data issues and anomalies • Ability to measure KPIs • Ability to be extended, customized, and rules driven • Ability to integrate with metadata • Ability to run as part of pipeline • Ability to validate accuracy, consistency, completeness, etc. • Ability to fix data gaps	• AWS Glue / EMR	• Informatica • Talend
Exploration	• Ability to bring own data for analytics • Ability to document experiments • Ability to have approval mechanism for promotion • Ability to work with various personas • Ability to collaborate • Ability to lay foundation for data science	• Databricks • Sagemaker • Jupyter on EMR	• Databricks • Paxata • Jupyter on Ec2/EMR

(*continued*)

Table 4-5. (*continued*)

Requirement	Key Features	AWS/Partner Tools	Third-Party Example Tools
Data Provision	• Ability to contextualize data • Ability to define and implement industry use case on data • Ability to integrate with security, RBAC, ABAC, etc. • Ability to expose data for multiple consumption pattern • Ability to industrialize data for data warehouse provision pattern • Ability to industrialize data for building hubs, marts, pools • Ability to industrialize data for search-based use cases • Ability to industrialize data for building Genome • Ability to industrialize data for building time-series reporting • Ability to industrialize data for point-in-time search	• Redshift • Snowflake • Athena • RDS • DynamoDB • Neptune • Elasticsearch	• Cassandra • Neo4j/DSE Graphs • Oracle • HBase • Solr/ Elasticsearch

(continued)

Table 4-5. (*continued*)

Requirement	Key Features	AWS/Partner Tools	Third-Party Example Tools
Data Consumption	• Ability to promote data to different personas • Ability to support ad hoc, long-running and complex processes • Ability to integrate with machine and artificial intelligence • Ability to handle multiple (hundreds) of connections concurrently • Ability to cater to multiple downstream apps with different protocols • Ability to handle multiple integration patterns for data access	• Quicksight • API Gateway • Lambda • RDS • Redshift	• Tableau • Talend • Informatica • D3JS • Kafka

(*continued*)

Table 4-5. (*continued*)

Requirement	Key Features	AWS/Partner Tools	Third-Party Example Tools
Data Science	• Ability to explore data and experiment independently • Ability to document experiments • Ability to manage models and approval of models • Have reusable feature management, normalization capabilities • Have reusable and config-driven models • Have integration with metadata layers • Have versioning of algorithms, data, parameters etc. • Have model approval and versioning system • Have feedback and model tracking capability • Have documentation for model reproducibility	• Databricks • Sagemaker	• DataIQ • Datarobot

(*continued*)

Table 4-5. (*continued*)

Requirement	Key Features	AWS/Partner Tools	Third-Party Example Tools
Data Governance	• Ability to onboard and register new data source • Ability to handle pre-ingestion • Ability to manage workflows and approvals • Design right zones and access policies • Ability to integrate with data management activities • Ability to allow for business glossary, data labeling, etc. • Ability to allow business users to browse data catalog • Ability to handle config generation and pipeline automation • Should be tangible and tie processes and people using technology	• Glue Catalog	• Alation • Collibra • ASG

(*continued*)

Table 4-5. (continued)

Requirement	Key Features	AWS/Partner Tools	Third-Party Example Tools
Data Security	• Ability to enforce perimeter, data, network, and access security controls • Ability to integrate with RBAC, ABAC • Ability to integrate with alert monitoring systems • Ability to manage ACLs, policies, and data at motion and rest processes • Ability to handle encryption, etc. • Ability to securely promote data and integrate between various systems	• IAM • Bucket Policies • ACL's • AWS Encryption	• Ranger • Sentry • Blue Talon • Kerberos • AD Integration
Alert and Monitoring	• Ability to keep track of processes and report issues • Ability to inform in case of error • Log management • Activity logging • Audit tracking	• CloudWatch • Cloud Trail • Lambda • SNS	• ELK • Dome9

Let's focus on some of the key topics mentioned and look at how we implemented each component for the customer.

- **Data Governance**

    "Data governance should not be an afterthought; rather, every project should implement a governance-first approach when building a data strategy." This is my mantra for defining and delivering any data strategy project. The question then comes to mind is, what is data governance?

    Data governance (specifically with respect to data lakes) involves building a synergy between people (who will build, manage, and use the data lake), process (such as the security, network, access control, data quality policies, data standardization policies, SLAs, etc.) that bind the usage of data in the data lake, and technology (which provides the enabler for the people and process). Figure 4-7 is a quick snapshot of how the governance process can be enabled.

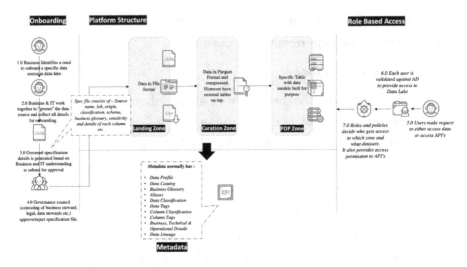

Figure 4-7. Simplified governance view

This process describes a way we proposed governance to this customer. This encapsulates the whole process described in the previous chapters and puts it into a tangible deployment strategy. We start by onboarding data sources into the data lake (sometimes described as *pre-ingestion*) to register and capture all the information for auditability purposes. Initially once the business user identifies a new data source to be onboarded, business and IT can work together to "groom" the data source, which includes identifying the source system details, ownership, frequency, type of data, etc. Once the details are recorded, they are shared and sent to the governance council to be reviewed and validated. The governance council is a group of stakeholders (from business, legal, etc.) who can validate and ensure the data source is legally, technically, and business-wise justified. Once the registration process is complete, the data pipelines can execute and move data across hops. This includes enriching the data, confirming the quality, capturing the metadata (with tagging, etc.), and building the central catalog and marketplace. Finally, the governance process overlays the security and access control plane on the data plane itself to provide the right kind of data accessibility and controls.

From these descriptions, we can define data governance as a collection of the processes (through technology) in Table 4-6 to enable specific personas (people) within the organization.

Table 4-6. *What Can Be Governed at Each Stage*

Category	Governance Assessment Areas
Strategy and operating model	Strategy and mission
	People and team structure
	Policies, process, and operating model
	Tools and technology
Data foundation	Business glossary
	Data lineage
	Data access and ownership
	Data privacy and security
	Data quality and trust
	Metadata management
	Master data management

(*continued*)

Table 4-6. (*continued*)

Category	Governance Assessment Areas
Data stewardship	Stewardship
	Project management
	Change management and user adoption
Governance performance	Performance management
	Productivity management

- **Data Ingestion**

  Data ingestion is the process of capturing and "hydrating" the data lake with data from internal or external sources. Figure 4-8 provides a quick reference to what the nuances of data ingestion are. Ideally, the ingestion workloads can be either batch, micro-batch, or streaming. Once that is established, the pattern of getting data through those workloads will differ based on Figure 4-9. Additionally, the data that is brought in can be of multiple types, formats, and sizes, and these data types bring in additional challenges in terms of conversion, flattening, formatting, compressing, etc.

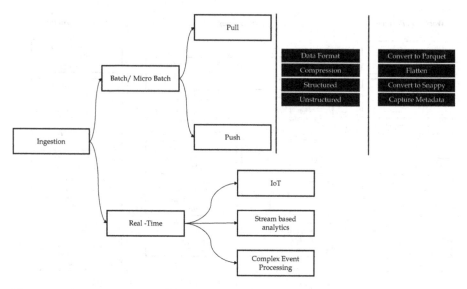

Figure 4-8. *Overview of ingestion process*

The following are some of the considerations
while selecting the right ingestion mechanisms. I
divided the whole ingestion selection process based
on workloads, capabilities it offers, and leading
practices. If we look closely, most of these practices
are easily managed by the AWS cloud providers;
hence , for this customer, choosing a cloud-native
solution was easy. See Figure 4-9.

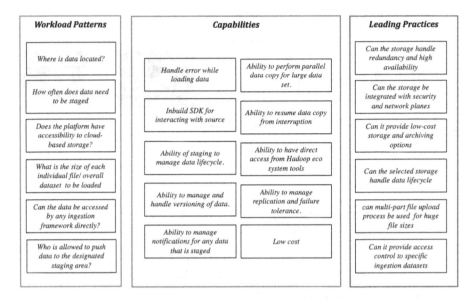

Figure 4-9. *Overview of workload patterns*

For our customer, we selected a Spark-based, AWS-centric, Glue-based solution for all batch loads and selected Lambda-based, micro-batch loads. There were no real streaming needs for the customer yet. However, from my experience, I have seen that if we scratch the surface enough and ask the right questions, typically 99 percent of workloads will be micro-batches instead of streaming workloads.

The following are some of the questions you can ask your customers if they are on the fence for streaming versus micro-batches:

- Does your company have the infrastructure set up to manage streaming pipelines?

- Does your customer maintain any other streaming
  pipelines? If not, then the inertia to start the first
  one is usually incredibly high.

- Do your stakeholders really need "real-time"
  data? A lot of the time real-time ingestion does not
  co-relate to real-time use of data for any business
  purpose.

- Is there any incremental benefit that you get from
  streaming that micro-batches don't provide?

- Does micro-batching (5 to 15 minutes or even
  event-driven) satisfy the latency requirements?

Figure 4-10 and Figure 4-11 describe what we
deployed for the customer. One is for a micro-batch
that was more event-driven with low latency and
low throughput versus other batch processes that
had high latency (running once per day) but very
high throughput.

The micro-batch was written entirely on Lambdas
(AWS). The idea of the design was to keep things
very modular so that it would be easy to extend,
expand, and customize. Figure 4-10 provides how
the process looked.

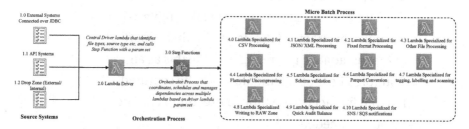

Figure 4-10. *AWS-based micro-batch view*

Figure 4-11 shows what we delivered as part of the batch system, which was doing the bulk of the work. We scheduled more than 200 pipelines fetching data from 200 different sources with multiple data and file formats across multiple source locations. The whole code was written in a very modular format in Python Spark and deployed as a package on AWS Glue for very large-scale processing.

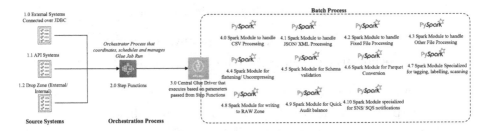

Figure 4-11. *AWS-based batch view*

- **Metadata Management**

    As discussed in earlier chapters, the process of *metadata management* is very tightly coupled with the ingestion process. Metadata should be captured as an "in-line" and active process during the ingestion pipeline execution. As part of this project, we built the profile and catalog of the datasets loaded into the data lake. Table 4-7 is a quick overview of the details loaded as part of the metadata capture. Again, take this as an example; the metrics and details to be captured have an impact on the time of execution and the resources needed.

Table 4-7. *Reference of What Metadata to Capture in the Data Lake*

Type	Subtype	Functions	Value Add
Business metadata	Catalog information	• Header inference • Schema inference • Hierarchies and folders • Entities (entity extraction: org, person, event, company, location) • Subject areas	• Define business policy • Grouping data based on business tags • Infer meaning hidden inside data • Understand business relationship • Drive governance policies • Define business compliance rules
	Business information	• Business terms • Business rules • Acronym, synonym, legal name, etc. • Business definitions • Semantic relationships • Data governance rules or policies • Sensitive data identification (PII/PHI/CPI, etc.) • Stakeholders (data steward/owner information)	

(*continued*)

Table 4-7. (*continued*)

Type	Subtype	Functions	Value Add
Technical metadata	Profiling information	Entity-level profiling • Count • Null count • Max • Min • Mean • Header information • Standard deviation • Ordinal position • Percent count • Blank count • Allowed values • Pattern count • Uniqueness • Length • Primary key candidate Attribute-level profiling • Frequency count • Frequency percentage • Least frequently used • Most frequently used • Compressions ratio and types • Formats	• Helps to define data dictionary for downstream systems • Helps to define data dictionary for building data models • Helps to define data architecture (logical and physical) • Helps to check data quality and data distribution (impact analysis) • Helps to identify correlation between different datasets and columns to identify relations between, which helps to get a holistic view of data (360) • Helps to design indexes and keys when converting the data into tabular format in the curation phase
	Structural information	• Source of data • Target of data • IT KPA • Tools used	

(*continued*)

Table 4-7. (*continued*)

Type	Subtype	Functions	Value Add
Operational metadata	Lineage Data life cycle Audit Executable metrics	• Job runs • Last time executed • Frequency of run • Status of run • Duplicate rows in each batch • Unique count in each batch	• Helps to know how data has changed over time • Helps to know who are assessing the data • Helps to know job runs/error during runs/ frequency or run to build KPI which can feed IT team for monitoring • Helps to create KPI on data load and data usage statistics for error tracking • Helps to drive security rules

Typically, the metadata captured is stored centrally. For most of the projects I have implemented, I have seen third-party products like Alation and Collibra being used. In my experience, I have implemented multiple Alation projects and have found their integration, API layer, and support model better suited. In this project, I recommended it, and we ended up using Alation as the third-party catalog and governance solution.

Figure 4-12 provides a quick reference on how overall the integration and pieces fit together (for the ingestion and metadata capture processes).

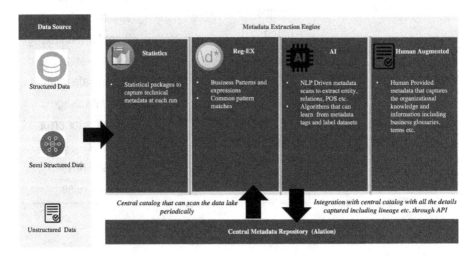

Figure 4-12. *Implementation view of the metadata layer of the data lake*

- **Data Curation**

  Data curation is the process of cleansing and standardizing the data to convert the raw data into a "single source of truth." As mentioned, this and data quality provide the two most important pillars in terms of data usability across the organization and building a robust data foundation.

  Figure 4-13 shows what we built for the customer. We created a very template-driven, low-code, and configuration-based framework that was able to handle the following rule types and executions. All of the following were developed through Py-Spark for batch and Python for micro-batch (AWS Lambda).

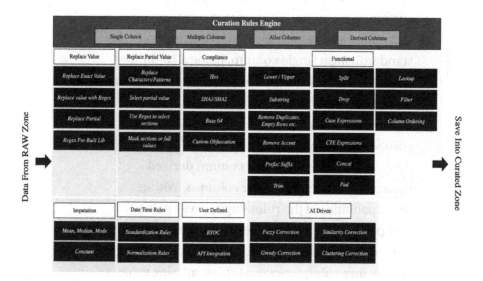

Figure 4-13. *Rules engine capabilities view*

The critical piece to Figure 4-13 is the modularity and compartmentalization of the code to ensure that we can keep extending it and adding more features based on additional use cases and the maturity of the data lake. It was important that we kept our eye on the overall prize and ensured we did not go down the rabbit hole of building "everything" as a "product" when the expectation was to deliver tangible business outcomes. We also ensured that everything was designed into a very modular (easy to extend, expand, and customize) solution that we could keep growing as our customer and their business grows.

The figure also provides a reference to the kind of standardization that was expected. For this customer, we built the automated rules engine that comprised capabilities such as fixing and standardizing a single column, fixing a column based on values from different columns, fixing data for a group of columns that share a common pattern or alias name (like date/time-tagged fields can correspond to order date, purchase date, shipping date etc.), or creating a new column derived from a combination of other columns. Within the capabilities of the rules engine, we enabled groups of policies that share a common pattern like regular expression rules, string manipulation rules, compliance rules, functional rules, missing value rules, etc. Each of these patterns then has a series of rules that can be enabled for a given data pipeline like AI-driven rules, etc., which have capabilities to autocorrect datasets through some fuzzy logic, greedy correction, or some unsupervised learning models (like k-means or k-nearest neighbors).

Similarly, functional patterns had rules such as trimming column values, splitting a column into multiple columns, performing lookup to standardize or enrich data, filtering certain records, enabling complex "create table as" expressions, removing accents, etc. The objective of the rules engine was to introduce repeatability and "low-code" options so that a data pipeline could choose a series of prebuilt rules to cleanse data as it moves from the Raw to Curated zones. The low-code option was specifically requested by our customer in this case as the goal was to have a very quick turnaround for implementing hundreds of data pipelines.

Finally, when we built this rules engine, it was important to focus on automation. By enabling automation, we improved maintainability, managed complexity, introduced flexibility (through modular approach), and enabled reusability to the whole development and implementation phase of the project. See Figure 4-14.

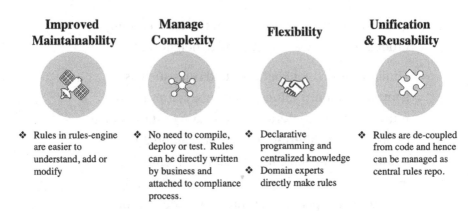

Improved Maintainability	Manage Complexity	Flexibility	Unification & Reusability
❖ Rules in rules-engine are easier to understand, add or modify	❖ No need to compile, deploy or test. Rules can be directly written by business and attached to compliance process.	❖ Declarative programming and centralized knowledge ❖ Domain experts directly make rules	❖ Rules are de-coupled from code and hence can be managed as central rules repo.

Figure 4-14. *Reference of need for a framework-based data lake design and implementation*

- **Data Quality**

  Data quality is critical to building a trustworthy data flow. There are lots of data quality tools available on the market. However, the biggest challenge is to ensure the data quality is "in line with" and becomes part of integral data pipeline flow. Most of the tools in the market can run data quality checks after data has already been used by downstream business processes only to realize issues in the data that lead to untrustworthy outcomes. The customer in this case was very clear that we needed to build the data quality module native to AWS so that it could be leveraged by all the pipelines we created, and additionally it needed to be inline so that only verified and validated data would go to downstream business processes.

  The data quality as a process needed to be "active" and "inline" and could not be an offline "one-off" process. Data quality needed to be spread across the supply chain of data management steps.

Figure 4-15 shows the data quality processes and how they need to be applied within the context of the data lake.

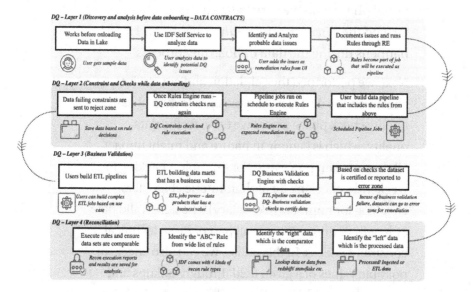

Figure 4-15. *Reference view of a data quality framework*

Figure 4-15 provides a glimpse of what we built for the customer. It talks about a four-step data quality process (the four pillars of data quality) where we identified the data quality checks into multiple phases of the data life cycle.

First, we introduced the concept of data profiling and the ability to inline fix the data through known rules and processes (we used our rules engine from the previous step) to help fix data before rejecting or flagging data issues. In an actual data platform project, the development process always runs the data quality checks in lower environments to

109

identify known issues and document data validation challenges that can be fixed before promoting the processes in the production environment. The initial phase where we try to fix known issues is termed the *discovery and analysis phase*, and it happens during the data onboarding process.

Next, we performed the validation and constraint checks on the data. This is typically a technical data quality check phase where we try to fix the data quality, which is generic in nature and might impact all downstream processes. For example, checking for weekends, checking for alphanumeric characters, checking for numbers (like salary), checking valid date or length or range, etc., are all part of these technical constraint checks. In this phase, we can also check for possible values of a given field through lookups and master data validation checks. The idea of this phase is to ensure that data that comes out of this process is "pristine" and "trustworthy" in nature and can be used for any downstream business processes.

Next, we included the business validation process. This is critical and at the same time contextual in nature. For example, we can test and check that once we build a data product (data mart for reporting, etc.), we can validate the reporting data and ensure it is business-wise acceptable. An example can be to ensure that once we join the product dataset and order dataset (as an example), we can ensure that the product revenue cannot be negative when the number of products ordered is more than 1. Within the context of our customer, this data quality step (business validation) is very important as many of the project requirements should have trustworthy business reporting metrics.

Last was the audit and reconciliation checks. These checks ensure that as we move data from external systems all the way to different zones and marts, we are not losing the value of the data either due to some network or due to precision or adjustment, etc., activities. Typically, the audits and reconciliations processes are not executed at the end but are sprinkled across the data pipelines as we continuously check the counts and sum, or dollar amounts, etc., to ensure we are not losing value of data anywhere in the pipeline.

This four-step data quality process ensured that we ended up with a well-rounded and well-oiled DQ framework for managing the data quality expectations for all pipelines running in the platform. See Figure 4-16.

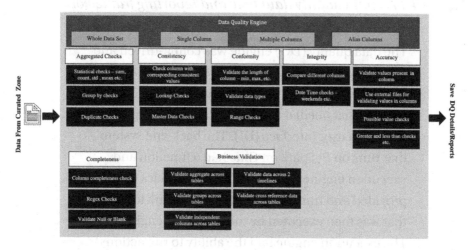

Figure 4-16. *Capability view of a data quality framework*

Figure 4-16 provides an understanding of the framework we developed for the customer centered around the previous data quality rules. We delivered the framework for our customer as part of the engagement, and over time this framework matured to be an integral part of our customer's enterprise-wide data quality and data management asset.

Figure 4-17 also provides context on how this overall process fits into the data pipeline.

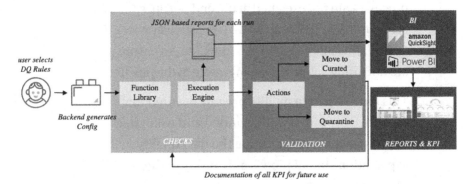

Figure 4-17. *Data quality data flow and reporting framework*

The data quality (DQ) engine was config-driven and was executed on the AWS native serverless stack using AWS Glue. The users selected the data quality policies that needed to be executed for each pipe through a user interface (UI). The back-end system that runs on PySpark on Glue had a function and execution engine that picked up the policies users provided and translated them into PySpark logic that was then executed by the execution engine. The execution engine had the ability to run actions on the policies executed. The action can be either to do nothing and propagate the data into the next phase (curated zone) or to identify the impacted rows (or the whole dataset based on what was the policy) and move those records into the quarantine zone (optional zone in addition to Bronze, Silver, and Gold, which is needed if we decide to isolate the incorrect records separately) and notify the correct stakeholders. Either way, the data quality engine was supposed to generate KPI reports and metrics that were saved for comparison and to keep a tab of the data quality issues and challenges the customers were facing on a day-to-day basis.

- **Data Products**

This is a critical piece to building business-centric applications. The idea is to use all the earlier processes such as ingestion, data curation, data quality, governance, metadata, etc., and reach a point where we can deliver use cases and contextual solutions to our customer's business needs. The stages before data product creation centers around building a "single source of truth" and focuses on getting high-quality data before the data products section uses the cleansed data for implementing it.

The data product building process is mainly using simple to complex extract, transform, and load (ETL) processes to build either a data mart or a data vault, etc. There are typically many ways to design and implement the data product such as building a wide table, snowflake, star schema, data vault, or the newer concept of a data mesh. See Figure 4-18.

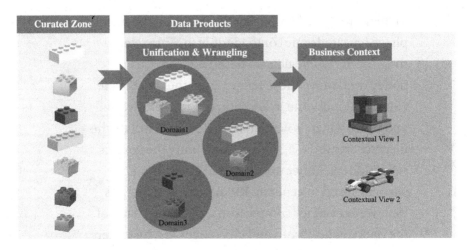

Figure 4-18. *Simplified view of data product implementation*

Figure 4-18 provides a quick visual representation of the process of creating a "fit-for-purpose" business use case–driven data product from curated datasets. Here, the "Lego" analogy was used to represent the fact that data in a Curated zone is well organized and well managed, but it does not have any business value attached. To build a data product, we can start by unifying the data into a domain layer that helps to identify which datasets can be assembled together, and finally the business context (or ETL) is added to the domain pieces to produce the outcomes that have a business value attached. As the next steps, let's take a quick look at some of these data product creation patterns and discuss how we implemented data products for our customers. This was a very powerful and representational picture that we showcased to our customer, and we show it here so you can have a better understanding of the process.

Wide tables: A wide table is possibly the most widely used data structure for building data products within the context of a data lake. The idea is to use "curated" data from the Silver zone and implement the business know-how as a SQL (or Spark) ETL script that can join and manipulate the data to build a big giant and wide (all columns in a single data structure) table structure for usage. This pattern was used for more than 80 percent of the use cases for this customer mainly because of the advantages it provides.

First, the ETL processes are used to create the data products to make it more visible and usable to the stakeholders, so typically the data products can power a business intelligence (BI) report or dashboard or some machine learning model. If the application that needs to consume the data product also needs to perform another round of data wrangling or joins, then it defeats the purpose of building a quick and usable data product. So, if the data product is already structured in a simple single table, then it helps to be consumed in a quick and easy way. The same story goes for machine learning models. All ML frameworks expect the data to be ready for feature engineering, which means the data needs to be organized in a single tabular structure where we can map the signals to the target value. This concept of wide tables is easy and is use case specific and contextual. The only downside to this is the reusability of data products to power multiple dashboards. These are some issues addressed by the next kinds of data product design principles.

Star and snowflake schema: Star and snowflake modeling is used beyond the concepts of data lakes. Hence, I will not be covering those in this book specifically. However, let's take a look into how the star and snowflake schemas can be applied to the concepts of a data lake and building data products.

The idea of the star and snowflake schema is to break down the datasets into facts and dimensions. Facts are measures of attributes captured such as sales price, amount, quantity, orders, etc. Dimensions are characteristics about the fact such as who bought the item, which date the item was purchased, what are the product specifications of the item, etc. If we draw out the facts and dimensions in a piece of paper, we will see that the fact is a table that can be represented in the center surrounded by dimensions (hence it looks like a star). Now, taking that philosophy to the data lakes, one way to design data products can be to break down the business use cases into facts and dimensions. For example, if we want to build a data product of monthly sales, then we can create a fact table called sales (that contains the sales details, etc.) and then extract the dimensions of sales (like information about the order itself, information about the customer who purchased the item, information about when the item was purchased, information about the employee who sold the item, etc.) into different dimension tables surrounding the fact table (see Figure 4-19). The idea of building a data product in this way ensures that we can use the customer or employee dimension to quickly build another data product that has a different fact (like an order report instead of sales report) and that we can ensure that data refreshes to the dimensions and facts can be applied differently. However, this comes with its own disadvantages. For example, the BI application that needs to represent the data product as a report or dashboard now needs to know how to join the facts and dimensions together to represent the exact use case. Snowflake and star schema modeling are common and widely used in the enterprise. For this customer, we ended up using less star schema modeling (as we build a lot of wide tables); however, for other projects, building a star schema was very common.

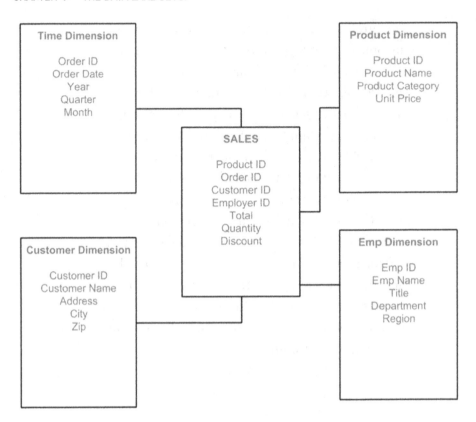

Figure 4-19. *Sample star schema example*

Data vault: Data vaults are logically and naturally
an extension of the star schema modeling process
to provide more flexibility in terms of changes and
adaptation. The main motivation of building a "data
vault" data model is to have flexibility and agility
with the ever-changing dimensions of the source
systems and need for accommodating changes to
business use cases. Let's start the explanation using
an example use case.

In the example in Figure 4-20 (and compared to the earlier star schema), the datasets can be divided into entities called *hubs* and *satellites* and then connected using *links*. We typically set up hubs as "entities of interest" that contain specific business keys and other related information. Links are entities that can record a "fact" or "transaction." They contain information of the hubs they belong to and typically are connected to one or more hubs. The satellite tables are used to capture a snapshot in time based on when certain events that can occur, etc., and typically are connected to a hub as it contains "dimensional data" in a granular level. In this example, the customer data is documented as a hub that contains the customer ID to uniquely identify it. The other details of the customer (such as demographic details and communication details) are modeled as two independent satellite tables and are connected to the right customer hub. Finally, we have a link table that connects two hub tables called *customers* and *orders* that contain the integration details.

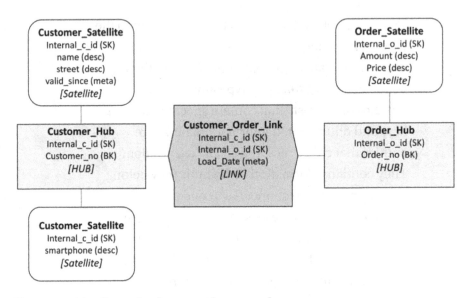

Figure 4-20. *Sample data vault example*

Typically, projects start with a wide table model and eventually mature to a data vault construct. The only downside to the data vault setup is the overhead and complexity to manage and maintain this compared to the wide table construct. In lots of projects I was involved in, I saw the trend where we started with the wide table and a business-specific data model and then eventually matured to a star schema and then to a data vault.

We talked about what a data vault is and also mentioned the maturity curve, etc. Let's now focus on how to extend the data pipelines to automatically "hydrate" the links, hubs, and satellite tables. This is something we implemented for this customer by implementing the logic through our PySpark/AWS Glue-based framework as a distributed process to handle a large volume of data.

The ETL processes we created for this has three parallel processes. The first one was supposed to update the hubs, and the responsibility of that process (ETL job) was to connect to the curated datasets, identify business keys (from previous loads), and then update/append/delete from the hub to keep the data consistent.

The next process for ETL was to hydrate the satellite tables; this process is comparatively easy as we just need to fetch data from the curated zone and append the satellite tables based on the business keys from previous loads. The only additional step was to generate some surrogate keys and map the business keys to the surrogate keys to connect the satellite to the hub.

- Finally, we had the third process (ETL) to create the link tables. In this process, we load data from the curated zone, get the business keys, identify the surrogate keys through a hub lookup, and then delete and insert the latest data based on the existence of the records. See Figure 4-21.

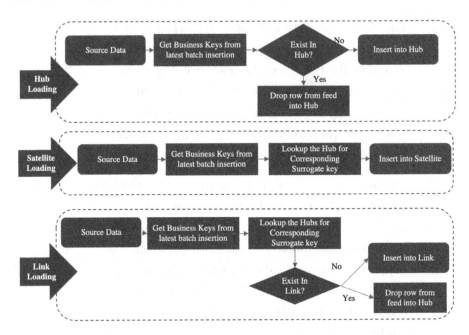

Figure 4-21. Data vault process flow for inserts and updates

These pipelines need to be integrated and orchestrated with the existing data pipelines, and hence the dependency and availability of datasets, etc., need to happen independently.

Data mesh: This is probably the most exciting piece of technology for this year. It is a new concept within the context of a data lake. I interacted with multiple architects to understand what the concept of a data vault means to them, and honestly, I did not get two similar answers from a group of ten. However, I ended up setting up multiple design sessions and invited multiple architects within the industry and technology domain to create a version (or at least my version) of a data mesh that everyone in that group agreed to and that was tangible and deployable.

First, let me define what a data mesh can be and what challenges it can potentially solve.

- Ideally, data lakes are built for the entire enterprise and often have multiple organizational units that want to own and manage their own data products (not only ownership of data but also the ownership of the processes).

- Typically, a central IT team owns the processes in the data lake, whereas the data still might be owned and managed (through conformity, etc.) by the organizational units (like the Finance team and HR team). Sometimes having a central IT team becomes a bottleneck specifically when the projects are managed on independent AWS accounts (that are owned and managed by the organizational units).

- Data sensitivity plays a critical role. Datasets loaded by the Finance team (for example) can have sensitive information that only a finance data scientist or finance data analyst might access; hence, managing these data assets (data products) by the individual organization makes it effective and secure.

- The whole concept of a data mesh works around the principles of "federated" ownership of data and its processes but still keeps a centralized governance to ensure every business owner conforms to a given set of organizational guardrails.

- The "incentive" of "ownership" of data products by the business organizations is done through "carrots" and "sticks." The "carrots" are the capabilities and possibilities it presents to the business domain teams to be independent, not be reliant on central IT and have their own authority of the business outcomes. The "carrots" are also the "resource and fund" allocations received from the enterprise. The "sticks," however, are the constant governance and security scans that are needed, and most significant is the responsibility of the business teams to onboard and manage their own resource and skill pools.

Taking the previous rationale into consideration, we defined what a data lake should be and how it needs to be extended to fit into the data mesh principles as follows. The idea is to organize data into a data lake construct (the typical Raw, Curated, Provisioned, etc.) and then design the data products into a highly scalable data warehouse that can support SQL-centric downstream applications. To achieve this in a truely decentralized way, we must ensure that we have right controls, right teams, and right skill sets to operate and manage the data products as a data mesh. The concept of a data mesh is to allow individual functional or subject-matter expert groups to create their own data products for their own needs but also share those with the right access and controls to any other groups who might need the value of the data product. In that way, the data products become true products within the organization, and they have their own life cycle. There are dedicated teams that manage the data products, but also they publish information (through a catalog of data marketplace) to inform other teams to potentially use the data product if needed. As shown in Figure 4-22, once we have created a sustainable data lake, we can build a data warehouse to manage the data products, but these data products can be owned by business units (enabling proper management, maintenance, hydration, and issue management) and eventually shared through common architecture principles like the hub-and-spoke model, centralized data marketplace, centralized access control, etc.

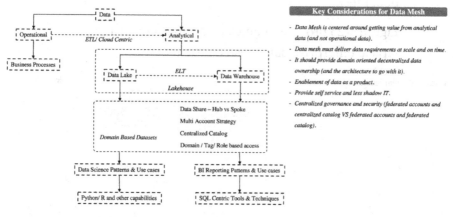

Figure 4-22. *Overview of data mesh*

The biggest challenge for an organization to implement a data mesh is the availability of the right resources, skills, and ownership. Typically for any organization, they have a central IT team that is more technology focused, and the business units typically have domain expertise. Now, for the business units to own and manage their own data product, they also need to have a proper IT and technical capability to ensure the right management and maintenance. This is why the idea of building a true data mesh becomes a challenge for lots of organizations. In my scope of work, I have seen customers who start with a central IT department and over a period of time evolve into the working dynamics of a data mesh for some of the business units. It is the simple philosophy of "crawl, walk, and run" to stand up an operational data lake, ensure business and IT are collaborating together to bring business facing and useful solutions, and then eventually mature toward the construct of a data mesh.

Figure 4-23 provides a quick understanding of a typical enterprise data mesh construct.

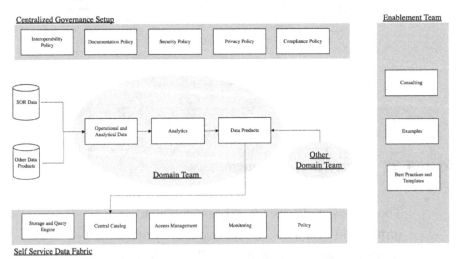

Figure 4-23. *Data mesh holistic view in terms of federated setup*

Figure 4-23 defines the typical building blocks for implementing an enterprise-level data mesh.

First, data mesh enablement should focus on defining and establishing centralized governance standards through interoperable policies, documentation, security measures, privacy policies, and compliance policies. The governance policies are regulated and moderated centrally but executed through federated processes.

Second, there should be a centralized self-service data fabric team that ensures common frameworks, patterns, building blocks, and design patterns around storage and job execution, centralized catalog management, access management, monitoring of jobs and processes, and policy (access and data privacy) across all business organizations. This will ensure uniformity in the technology stack, as well as reusability of common patterns and modules to make new data product creation and maintenance easy and manageable.

Third, there should be an enablement team that consists of specialized skilled resources with engineering and domain expertise that can help facilitate the organizational units to get up to speed and be the subject-matter experts (SMEs) through example sharing, templates, documentation, and consulting. This will enable business and organizational units to worry less about resource and skill mapping all the time.

Finally, the business organizational units or domain teams that will use data mesh as a platform to enable their own business processes and create data products. The teams will be responsible for using the enablement teams and playing along with the rules defined by the central governance and self-service teams to onboard their data, perform analytics, and expose insights as data products.

- Now, the journey to a self-contained data mesh setup is gradual and goes through multiple phases. No organization can align their operational processes to follow a data mesh setup overnight. Figure 4-24 shows the progression and evolution that is typical with most of the organizations that want to have a sustainable data mesh implementation.

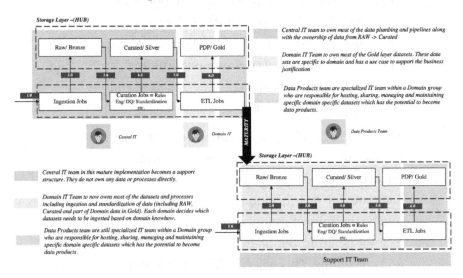

Figure 4-24. *Data mesh maturity and ownership view*

In Figure 4-24, the example organization starts with a typical data lake with the three zones we discussed. In this case, the ingestion jobs, curation jobs, and all the other data plumbing processes are managed and owned by central IT. Typically the business (domain) team can own some of the data products that are built through a wide table or star schema philosophy, and if the business unit has the IT capability, then some of these data products are managed by the business teams themselves (the ETL process and not the underlying platform, etc.). They can start with a small IT team in-house for the business domain to get accustomed to how the data lake works.

Next in the maturity curve is when the business units have their own dedicated IT team and can take over their own process ownership (ingestion, curation, etc.) along with ownership of the data so that they are self-sufficient and independent. They manage and publish their data products into a central catalog, and any other teams that need access to these data products have to contact the owner data product team.

Finally, before wrapping up the data mesh discussion, one important question is how these data products are shared, including who pays for using a shared data product (the product creation team or the team requesting access). This question can be answered by certain technical capabilities that AWS as a platform supports. There are concepts of shared objects where a group can share (through RBAC and ABAC policies) the whole or a section of the data product with a requesting business unit through the separation of compute and storage so that the data is owned (read only) by the "owner" business unit but any analytics or queries that are executed on the data product by the "consumer" business units are executed in a compute layer that is owned and managed by the requesting business unit. This process is explained in Figure 4-25.

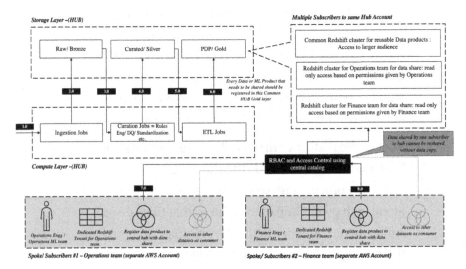

Figure 4-25. *Data mesh implementation view*

In this example, there is a central IT managed hub (data lake) that has the typical layers. There are two business units in the bottom (operations and finance). Ideally, they would like to "own" their own processes (ingestion, curation, DQ, ETL, etc.) along with owning the data. The data products created would be registered with a central catalog that is managed in the central hub. For example, if the operations team has created the data product and the finance team needs access, then the central access policy will be enabled in the hub account so that the Finance team can get access to the (whole or partial) data product after the operations team has granted access to the finance team. The finance team does not need to copy the data or refresh the data; the data product is shared "live" by the operations team to the finance team. However, any queries or ETL executed on the live data will still be running in the compute layer within the finance team's account. This kind of an access model ensures centralized data with decentralized access through federated accounts for clear ownership and clear separation of concerns.

Now that we have discussed the data product patterns in detail, let's focus on what it would take to build a comprehensive framework for ETL. Similar to the concepts of building the rules and data quality engine as explained in previous sections, we ended up building a very configuration-driven and repeatable framework for building data products through ETL. Figure 4-26 provides a quick reference on the capabilities of the framework.

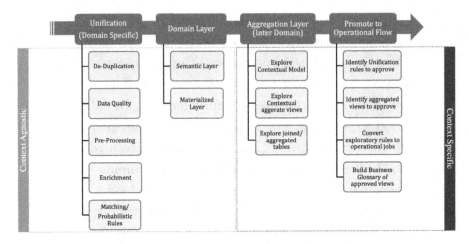

Figure 4-26. *Data mesh implementation through ETL process*

Figure 4-26 explains the four stages, starting with domain-specific unification that can help in performing preprocessing, matching, business validations, etc. The second layer is building the semantic or materialized layer that provides guidance on the data modeling techniques that we need to employ for the use case. The next two layers, the aggregation and operational flow, focuses on building the data product based on the use case through exploration and then adding the security (role-based access versus views, etc.) on the data model created.

- **Data Consumption Patterns**

    This section focuses on the process for providing access to the data products created in the previous step. For this current customer, we had the requirement to only enable the data products through business intelligence (BI) and reporting workloads. I will cover a few options here based on the work I did for the project; however, there will be multiple other consumption patterns that might be needed for other projects.

Figure 4-27 provides a quick reference to the kinds of consumption patterns that are common across the enterprise (with a brief description of the same) and then provide examples of how that consumption pattern can be accessed.

Consumption Pattern	Description	Data Access Patterns
BI/Reporting	Business Intelligence and reporting represent the standard traditional data consumption from data warehouses and databases. Typically, back-office IT powered, corporate executives and business leaders leverage this type of consumption for responsive dashboards and intelligent alerts.	• JDBC/ODBC • API/Microservices
Data Science/ Advanced Analytics	Advanced analytics provides experienced data engineers and data scientists access to digital data including operational, transactional and unstructured data sets from both internal and external sources. BI tools and machine learning libraries ensure that data exploration and discovery occur in the data pipeline.	• JDBC/ODBC • Pub/Sub • API/Microservices • Search • Broadcast
Machine Intelligence	Bringing in data from various sources into a single data storage that enables a far more efficient access pattern. With IoT and edge computing, devices through channel interfaces, have access to relevant streaming data. Modern flash-based data hubs that stores data from IoT and other sources act as that central data hub.	• JDBC/ODBC • Pub/Sub • API/Microservices • Search • Broadcast
Data Services/ Analytics Apps	The enterprise data services/marketplace is built for "Data Shoppers" seeking data for their analysis. Creating a faster, easier and efficient access to the data makes the intelligent enterprise to build a Data Services.	• JDBC/ODBC • Pub/Sub • API/Microservices • Search • Broadcast
Ad-hoc Analytics		• JDBC/ODBC • API/Microservices • Virtualization

Figure 4-27. *Data consumption patterns*

Figure 4-27 explains the five common consumption patterns that the enterprise data consumers want to interact with the data products: the BI reporting for dashboards, the data science team for advanced analytics, the ML engineering team for building the ML models, the analytics app users for data shopping, and finally the ad hoc analytics users for any interactive query capabilities. These personas need different tools and technologies to access the data products. The third column in Figure 4-27 talks

about the common patterns for each of the personas
(without naming their favorite tool) as how they
can access the data products. Connections through
JDBC/ODBC/APIs, etc., are very common across
tools of choice.

Additionally, Figure 4-28 provides a quick guide into
multiple consumption methods based on personas
and how each persona has a different need from the
platform itself and the need to start consuming the
data products.

Figure 4-28. *Data access patterns*

Figure 4-28 shows the kind of personas needed.
This includes executives who want to interact and
view a precanned report on a corporate level KPI,
business users who are a group of people and are
more focused on the operational KPIs and intuitive
outcomes, data workers who work on the data
products and can analyze them further, and data
scientists who can work with a combination of

curated and data products. The idea of Figure 4-28 is to connect the consumption patterns from Figure 4-27 to the personas and why they would need access to specific data products.

Next, Figure 4-29 talks about the ways in which these personas can connect (technically) to the data products generated and registered. Figure 4-29 shows the three-layer architecture that was enabled for the current customer.

The bottommost layer consists of the systems where data is stored. These storage layers can be legacy based or can be some of the modern data product platforms and patterns we have discussed. The middle layer discusses the technical way the systems can connect to the consumption patterns. This gives a 360-degree view of the people, technology, and process involved in the data consumption process. Figure 4-29 provides an indicative capability of how consumption patterns map to the data connectors to the data storage layers.

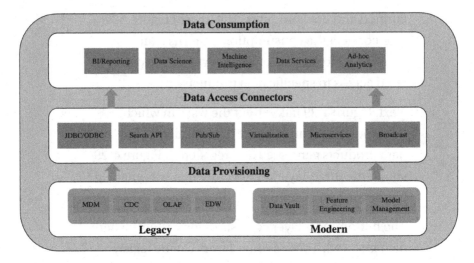

Figure 4-29. *Data access and consumption flow*

- **Data Protection and Compliance Through RBAC and ABAC**

  Although we did cover the RBAC and ABAC part in the security section, I feel we should spend a few moments here discussing how this data protection piece is typically integrated with the data flow. At this point, we have loaded data into the Raw and Curated zones and ensured that all data quality, standardization, and enrichment is done. We have not yet allowed interactive access to the Raw and/ or Curated zone to anyone. The data access was until now done through an automated process, which can be an ETL script or similar to build the data products. However, because of the nature of businesses, we now need to provide access to data (on a need-to-know basis based on the user role, etc.) to different personas (discussed earlier) from the Raw and Curated zones.

For this customer, we introduced the ABAC and RBAC using the AWS service called Lake Formation. Now, the general principles remain the same; however, the implementation might change depending on the choice of cloud vendor if you choose to implement something similar through another cloud provider.

So, this solution takes a bit from each of the previous sections discussed. We designed the metadata-driven data ingestion process for the customer. As a result of the data grooming process and capturing the metadata, we had the option to include the business glossary, tags, and business terms to the datasets and the columns associated. As a by-product of this well-engineered data intake and metadata management process, we ended up with a Curated zone that has all the data "well tagged" and "well classified." Once this prerequisite was established, adding the AWS Lake Formation and providing dynamic access policy was comparatively easy.

Let's start by defining what AWS Lake Formation service is. AWS Lake Formation is a serverless and managed service that makes it easy to secure and centrally govern data within S3 and Redshift. Now, as we built the data lake for our customer on S3 (Raw and Curated), we had the chance to work and implement AWS Lake Formation to help secure it. AWS Lake Formation simplifies security management through dynamic role and tag-based access. It also helps to provide and simplify user-based access to data from the lake through a regulated interface/service called Athena that is ideal (for our customer here) for all interactive access. AWS Lake Formation works through a central catalog called the AWS Glue catalog, which makes it easier for us to push and manage all the metadata that we captured into a central place. To make this easier to manage access and centralize the policies, etc., for this customer we stood up multiple AWS accounts as part of one single production setup. So, in other words, the customer's AWS production setup was logically a big AWS setup but physically consisted of multiple AWS accounts. One of the many accounts for this production setup was used for centrally managing AWS Lake Formation and all policies, permissions, and access criteria. See Figure 4-30.

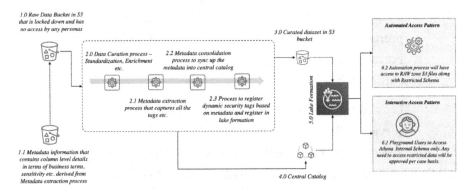

Figure 4-30. *Role-based and attribute-based access control for data products in a data mesh*

Figure 4-30 depicts the whole process. The data is available in the Raw zone initially, and then through the curation and metadata process we end up updating the Curated zone and the central catalog. Once the catalog and data are made available, a dedicated module creates dynamic policies and registers them for dynamic access to the data. The rightmost access patterns for interactive and automated access goes through Lake Formation and ensures they get access to only specific sections of data. Ideally, only interactive access is managed for RBAC and ABAC processes as the automated process runs for the whole of data. The interactive access request for data is governed, controlled, and managed through AWS Lake Formation to ensure the dynamic policies are managing the access to the data within the data lake.

In Figure 4-31, let's investigate the overall process map for Lake Formation and the dynamic access control processes. The setup starts with identifying and isolating one master Lake Formation account that can be used for all access control policy and rule setup. The account should create at least one dedicated admin role for Lake Formation usage. Once the admin setup is done, we need to ensure that permissions such as the ability to access Glue data catalog, ability to assume role into another account to grant access, ability to share resources with other accounts, etc., are permitted to this new admin role. Once these prerequisites are met, then the admin role can be used to provide some initial default grants and permission, etc. For this

particular client, we also ensured that the database
creation process is not ad hoc, and every database
has to be registered. Hence, in this case, every time a
new database is created, the Lake Formation admin
account has the full control to enable and grant
default permissions before others can start using
it. Once the database permissions are enabled, the
process enters the automation phase. In this phase,
every time a new table is created, it triggers this
lambda job that can use the lake formation APIs
to identify the sensitive columns, create dynamic
policies, assume roles, and assign specific tags to
columns. Once the tags are generated and the grants
are assigned, any user when trying to query the table
from the database will get access to only specific
columns. This was an innovative and automated way
in which we enabled our customer to automatically
manage the role and attribute-based access to large
volumes of data within the data lake.

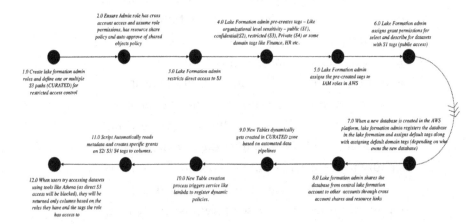

Figure 4-31. *AWS Lake Formation process for role based and
attribute based access control*

- **Data Reconciliation**

  We discussed the data reconciliation piece in the data quality section, but as our customer was very specific about data reconciliation, I want to discuss this topic again. The overall concept remains the same as discussed in the data quality section; however, I will go through some of the specific use cases related to the customer requirements here. See Figure 4-32.

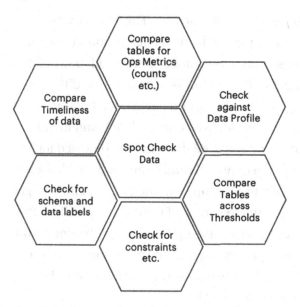

Figure 4-32. *Audit balance controls for reconciliation*

Figure 4-32 is just a recap of what we enabled for this specific customer. We enabled seven capabilities within the features of reconciliation as follows:

Check against profile: This is where we compared a baseline (from previous or established) pipeline as what a typical profile of data coming from a specific channel should be, and then we evaluated the new data profiles from that channel against the baseline. This was a true sense of detecting data drifts.

Compare tables across thresholds: This is more like a matrix comparison. We compared statistical measures against groups (like average salary per state) from baseline tables (raw or system or records) to target tables (Curated or Gold).

Compare timeliness of data: This feature was about comparing the data latency and informing downstream applications on the availability of data. Typically, the timeliness of data means different things for streaming and batch data. For this customer, we focused only on batch and micro-batch. We compared the modified date of data from the system of records and compared it with the ETL/data load timestamp from the data pipelines to get a view of the freshness of the data. Typically, if the source systems were updated every 30 minutes but we were processing the data at the end of the day, it meant that the data was typically a day old by the time we processed it.

Check for constraints: This is where we performed basic sanity; ideally it is covered in the data quality section, and we did not use this feature for the customer, but it would have been ideal to cover some well-known checks here such as ensuring the salary column is decimal and not string, etc.

Check for schema and data labels: This is more enforcement; we highly encouraged a governance-first and well-documented data onboarding and registration process. This reconciliation was to ensure that the schema and data labels are documented and compared against from source to target. This enforcement eventually helped build the central catalog and lineage.

Compare tables for ops metrics: This is an extension of the comparison against the table thresholds. This was a feature where we compared the data that landed into the data lake against previous load dates or data from previous runs or data from lookup master tables, etc.

Spot check of data: This was the simplest and most useful. We had ad hoc scripts where we compared (for example) dollar values of a certain column from Raw to Curated to Gold, etc., to ensure any precision, etc., are not getting dropped.

We can have other measures and features for reconciling data that other customers might use, but we found these measures (along with the four-step DQ framework) to be very robust and versatile for our reconciliation and validation (not only for this client but overall). In Figure 4-33, let's take a quick look into how this is connected to the overall data pipelines.

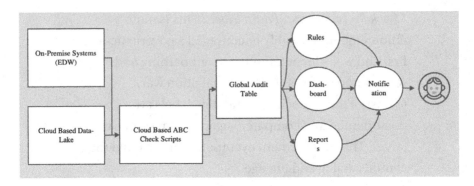

Figure 4-33. *Audit balance control process flow*

We created the reconciliation engine as a framework that has rules, measures and metrics. Once data starts coming into the lake through automated data pipelines, execute the reconciliation checks based on predefined rules to document, and take actions on the outcomes. Typical actions were informing the data owners, informing the data lake monitoring teams, etc.

- **Alert and Monitoring**

    This is the next logical capability within the data flow process. Data pipelines will fail, there will be production issues, getting a production bug or late-night calls are not exceptions...those are part of every delivery team's way of life. The success is measured in how quickly one can get the ball rolling in production again. Proper logging and monitoring go a long way to fixing issues and bringing stability to the chaos. As it is critical and important that all data pipelines are well governed, we have seen many levels of log and alerting mechanisms that provide the right level of control. See Figure 4-34.

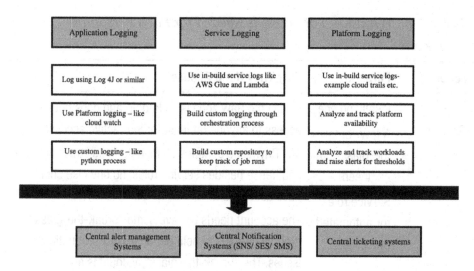

Figure 4-34. *Central logging and alerting process*

Figure 4-34 provides guidance on the alert monitoring systems. Typically, we enable three levels of logging: one from the application code level, one from the service on which the code is deployed (AWS services), and one from the platform level itself (AWS in this case). All of these logging capabilities are integrated further with the alerting features. For this customer, they already had a central notification system in place that had a proper channel of communication and alerting mechanism. We tapped into the same by integrating our alerts to the central notification process to use a company-wide alerting and notification process.

Setting the Right Access Control for Each Zone

We have talked about the right access control and personas who would get access to each zone in previous sections. However, as with every project, this becomes a point of debate, so Table 4-8 is a quick recap (as the same information is scattered throughout previous sections).

Table 4-8. *What Kind of Access Is Needed for Each Persona in Each Zone*

Zones	Personas	Why They Need Access
Raw (Bronze) zone	Ideally no one A few "break-the-glass" data engineers Service role for automated processes	Ideally the Raw zone is locked up and does not have access to anyone (at least for interactive access). The Raw zone contains all PII data in its "true" form, and hence besides service roles, no one else should have access to it. The account needs to have some "break-the-glass" production roles to help debug and fix production issues. The "break-the-glass" account is an on-demand basis and is provided for a specific period.
Curated (Silver) zone	Data engineers based on specific roles and access Data scientists based on specific roles and access Data analysts based on specific roles and access Service role for automated processes	The data in a curated zone typically goes through standardization, cleansing, and enrichment and hence is needed by analysts, data scientists, and engineers for building data and ML products. Access to data in the Curated zone should be based on the role the user has (like finance data scientist versus HR data scientist). Therefore, the data in the Curated zone is integrated with ABAC and RBAC controls based on roles and permissions. Typically, sensitive columns are obfuscated. The access to the Curated zone should be read only by any user (and can have write permission for service role which can process data from raw to curated without human intervention).

(continued)

Table 4-8. (*continued*)

Zones	Personas	Why They Need Access
Provisioned (Gold) ,	Data engineers based on domain Data scientists based on domain Data analysts based on domain Service role for automated processes	The main difference between the Curated and Provisioned zones is the organization of the data. Data in the Gold/Provisioned zone is business-driven and organized as data products (through a domain tier). Hence, access to the data sources here is the same as the Curated zone but with additional domain-based grants.

Understanding File Formats and Structures in Each Zone

I will cover this topic based on this specific need of the customer. Again, I think this is an optimal solution in most customer implementations, but I also have seen cases where other decisions were made based on the requirement, technical debt, choice of platform, etc. See Table 4-9.

Table 4-9. *What Kind of File Format Is Needed in Each Zone*

Zones	Data Format	Rationale
Raw (Bronze) zone	Ideally in the original format that data is available from the source. If data is pulled from source (from relational data sources, etc.), use CSV.	We want to keep data as close to the source as possible, and hence we want to keep a version of data that is not altered and in the same format as the source (which can include CSV, TSV, PSV, Fixed width, JSON, XML, PDF, etc.). Data from sources that do not have a format (like pulled from RDBMS) can be ideally in any format; however, we have seen that keeping it in CSV helps to view/validate/debug easier.

(continued)

Table 4-9. (*continued*)

Zones	Data Format	Rationale
Curated (Silver) zone	The preference is to have a common format. Typically the Parquet format is the most common. Of late a lot of customers are keen on new file formats like delta and iceberg. New data warehouse/lake house solution providers are enabling capabilities like Snowflake, etc., to be the common Curated zone.	Parquet is chosen in the Curated zone because of the systems that interact with the data here. Applications like AWS Glue and Databricks use Spark (the de facto standard of distributed processing), which can take advantage of the columnar file formats like Parquet. Additionally, Parquet can reduce cost (again due to being columnar) and integrates well with all data processing and analytics systems. Parquet has some limitations like being nonmutable and does not support SCDs; hence, we are seeing some new file formats (like delta) gaining popularity (which is again based on Parquet). Newer platforms like the snowflake have proprietary file formats and they advise usage of the same for the Curated zone.

(*continued*)

Table 4-9. (*continued*)

Zones	Data Format	Rationale
Provisioned (Gold) zone	Typically uses Parquet for enabling bulk loads and copy. The final copy of data does not have a file format (like snowflake/Redshift).	Typically, the Gold zone data is stored in a data warehouse. In that sense, data truly does not have a file format that needs to be explicitly managed; however, many scenarios need data to be enabled for bulk copy, and this is where Parquet (or a derivative of Parquet like a delta lake) provides high throughput.

This covers data lake structures and ways of working. This was the most important part and the heart of the whole data strategy project. Although it took the most amount of time to get it right, it was also the most creative and highly enjoyable part of the customer's data platform enablement journey.

Key Takeaways

To recap, in this chapter, we focused on building an end-to-end data lake. To achieve that, we broke down the entire solution into eight unique capabilities and started designing and implementing one at a time. By the end, we discussed the reference architecture, blueprint, example templates, and example outcomes that were used (and which you can use in your project as examples/templates). Now that we have the data lake up, our next step is to build the production playground.

CHAPTER 5

Production Playground

Objective: Production Playground

Imagine that we have a solid data strategy on top of a data lake. Once we have datasets loaded into curated tables, we must answer the most important question: how do we enable the data analysts, data scientists, and data transformation engineers to come up with "business-changing ideas" and implement business cases (data and ML use cases) that can add value to the organization?

In this phase, I focused on building a production playground setup that enables users to be onboarded (through proper governance and auditability) and get access to the "right" dataset so that they can build their experiments and evaluate business value before deciding on which experiments (from the list of active experiments) can be moved into production automation (in a true production setup).

There are two critical points here. First, a production playground is not a development environment. It provides "proof of business value through experimentation" rather than "proof of technology" (which needs to happen in lower environment). Second, this is called a *production playground* because we need to access production data to build use cases that delivers business value that business users can vet and approve.

Lastly, the production playground should not be a mock "production automation" setup where teams are creating shadow IT and setting up scheduled jobs, etc. The purpose of the playground is to create a business value and eventually get approval (and funding from business) to move the projects into an IT-managed "production automation" setup that can be continuously scheduled and monitored.

The Recommendations

As we have seen up to now, not all use cases are predefined and specified as the project starts. Most of the projects start with some basic understanding and predefined use cases. For others, it is important to provision a platform that is secure, governed, and managed and allow users the freedom to build "next best idea" that can open potential new use cases. However, to implement the "next best idea" that can revolutionize the business outcomes, it is important to work on production data and not just dummy data that cannot provide business justification and confidence.

This production playground should be an independent setup that has secure access to production data but also has guardrails, isolation, and separation of concerns that provide a safe zone for validating ideas. The ideas should mature over time and align with the business to identity the potential ones before packaging them as a deployable solution.

The use cases that we implemented for this customer were classified into two major categories. First was the well-documented, well-articulated, and well-defined set of requirements that can be implemented without doing a lot of research and experimentation. The second kind was the "unknown" use cases that needed a lot of research, experimentation, and acceptance before being able to finalize the outcome and tag it for production release. This overall classification (into two categories) is very obvious for enterprise-grade data lake projects because once we create a central data hub, the possibilities are endless. We should not only focus on what we know, but we should invest in identifying the "art of the possible."

This begs the question, what is the ideal place to build new work (create new data products or machine learning products)?

Before answering this question, let's understand what the exact need for this experimentation is and what issues we can solve with a production playground.

What Is a Production Playground?

Note the following:

- A production sandbox is an extension of a production data lake platform.

- It provides personas-based access to production data for quick analytics/prototyping with the security standards and practices of the production environment.

- As this is a production environment, personas like data scientists and analysts have access (based on roles, etc.) to live and timely data.

- A production sandbox has guardrails and controls in place where it provides a secure and user-based area for mixing production-grade data and interim use case results.

- This differs from other areas of the central data platform (or a data platform) in the sense that it will not allow scheduling, long-running processes, etc.

In other words, a production sandbox is an environment that allows data scientists and analysts to have access to live and timely data from the production data lake platform for quick analytics and prototyping. It is a secure and user-based area that follows the security standards and practices of the production environment but also has guardrails and controls to make sure that only prototyping and analytics on production-grade data is done. This differs from other data lake platforms, such as the CDP, in that it does not allow long-running processes or scheduling. This ensures that the use of the data is limited to prototyping and analytics and that the production data is not used for any other purposes.

What Issues Will This Address?

This will address the following issues:

- Long-standing issues of enabling data scientists, BI analysts, etc., to have a "self-service" way of building quick prototypes and experimenting on real live data for business value

- Quicker time to market for a mission and time-critical use case

- Long-standing issues of making live data available for quick analytics without having to make copy or share snapshots

- Ability to wrangle production data (read only) with "bring your own data" or "public data" for quick experimentation and prototyping

In other words, this is describing the importance of data accessibility and experimentation for data scientists, business intelligence analysts, and other professionals. The text is highlighting the need for a "self-service" way of quickly building prototypes and experimenting on real, live data. This would enable a quicker time to market for mission- and time-critical use cases. Additionally, this emphasizes the need for a way to access production data (read-only) for quick experimentation and prototyping so that professionals don't have to make copies or share snapshots of data.

This would allow professionals to quickly and easily access the data they need for their projects and to quickly experiment, prototype, and analyze data to find business value.

What Is a Production Playground *Not*?

Note the following:

- It is not a DEV/ UAT or any other lower environment.

- A production playground differs from a lower environment in terms of the value it brings. A production playground is used to prove a business theory instead of prove a technical capability.

- It is not a temporary account for one use cases.

- This is not an environment with less access control or lesser security standards. The user role and group will still determine who can access what.

- This is not shadow IT. No scheduling/automation will be allowed. Data will be purged on a regular basis to keep things clean.

What Does the Production Playground Consist Of?

Note the following:

- A dedicated zone (like Silver or Gold) that has its own dedicated compute and storage

- Granular access to project/use case–based storage area and service principals to allow project teams to have their own independent and "self-service" workspace

- Guardrails that enable sandbox users to have (role based) access to CDP Silver/Gold/Redshift data for read-only purposes (write is disabled)

- Ability to bring own data (BYOD) in the designated project workspace and ability to wrangle the same with read-only copy of data from Silver/Gold, etc.

Let's quickly take a look at what we defined earlier and how that relates to the concepts of a playground. Ideally, there are two kinds of personas who benefit from the playground construct. One is the advanced analytics user who needs to experiment and analyze data from the data lake to create data products. The other is the data science user who needs access to the data to build machine learning models as products. Although the two personas have somewhat similar expectations and requirements from the production playground area, we are currently focusing on the advanced analytics user for data product creation as part of this book. See Figure 5-1.

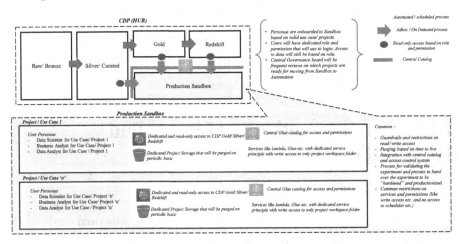

Figure 5-1. *Overview of the production sandbox and where it sits in the overall data lake*

The top-left section is the traditional data lake blueprint that we have been discussing the whole book. The data flows through Raw, Curated, and Gold and then can be registered as part of the data warehouse. The bottom-right section calls out the production playground. This is the area of interest for this section of the discussion.

Ideally, this production sandbox is an isolated yet logical production area that has access (through guardrails) to curated and provisioned zones of the data lake. The production sandbox area is managed by the central catalog and guardrails. The idea is to have users getting read-only access to specific datasets or sections of datasets based on the roles and project profiles. This production playground area is a specialized area and should not be confused with a development environment. The intention of this area is to prove a business value with actual data instead of delivering technical values (bug fixes, etc., which are more aligned to the dev environment). Because of this reason and rationale, the onboarding of users in the playground area is based on project needs and the role the person has within the organization (like data scientists and data analyst, etc.). The playground area is managed and governed by a series of guardrails and control mechanisms and has a strict security pillar as the users in this area have access to production data. Double-clicking into this area, one way to organize the production playground setup is by using a project-specific organization structure. This ensures that when a project is onboarded, a new Active Directory role is created that is assigned to all users in that project. The users then can have specialized access to certain domain and sensitive datasets based on the profile and policies.

Additionally, the person gets access to a special designated area to coordinate the work with other members of the group. This designated area is not accessible to other users or other project members. Similarly, the tools enabled in the production playground are always whitelisted and preconfigured (with security and guardrails). This is to ensure that users are not using this production playground area as their local workstation.

Lastly, this production playground area is controlled by a governance board that runs on regular cadence with the project teams to take stock of the work done and ensure frequent cleanups and purging of unused and managed projects. The intention of this production playground area is not to develop a "shadow IT" where projects live forever. Any project that originates here needs to be vetted and evaluated before either moving the project to the Gold area so that it can be managed by the central IT team or purging it if no value is found. The objective of this area is to provide a quick time to market for "next best" data and machine learning products to originate here (without the IT red tape) but eventually goes back to the automation process to be managed by central IT (or domain IT for a data mesh).

Key Takeaways

To recap, in this chapter, we took a step forward. We talked about the data lake setup where we "ingested" the right datasets and enabled users to get access to enterprise data. We focused on enabling users to prioritize and build their use cases that can provide business value.

This is a critical chapter that focuses on how to use the data lake and how to build an innovative culture of democratized data access and business value through experimentation. A lot of projects end their responsibility after onboarding the data into the data lake. Unless we focus on the production playground, the job is only half done!

CHAPTER 6

Production Operationalization

Objective: Production Operationalization

This is the next logical next step after the previous chapter. We have loaded the data and democratized the content. We then enabled a production playground and enabled different personas within the organization to build experiments that can provide business value. After a lot of these steps, we have our "next best idea" identified. The next best idea can be an ETL job, reporting job, machine learning pipelines, etc., which can help bring tremendous business value to the organization. So, the logical question is, how can we move the experiment in the production playground into a managed, scalable, and monitored production job that can be owned by central IT (versus the experiment team). Similarly, once the business has vetted the business outcome, how can the code get through the security and other processes to move into production?

This chapter will focus on answering these questions.

© Nayanjyoti Paul 2023
N. Paul, *Practical Implementation of a Data Lake*,
https://doi.org/10.1007/978-1-4842-9735-3_6

The Recommendations

In this chapter, we will focus on the aspect of moving the data products and processes we have created in lower environments or in the playground area into the automated production Provision zone (Gold zone). We will focus on continuous integration and deployment and how we enabled automation to move the code and assets into an IT-centric and IT-owned platform.

Typically, a general CI/ CD integration process has the following steps (within the context of a data lake):

- *Create a continuous integration (CI) pipeline*: This involves setting up a CI server to handle the automation of builds and deployments.

- *Configure the CI server*: This involves configuring the CI server to integrate with the data lake, such as setting up source control systems, source code management, and artifact repositories.

- *Set up automated tests*: This involves setting up automated tests to ensure the quality of the builds and deployments.

- *Deploy builds to the data lake*: This involves setting up automated deployments of the builds to the data lake.

- *Monitor the data lake for changes*: This involves monitoring the data lake for any changes or updates to the codebase and making any necessary adjustments.

- *Implement continuous delivery (CD) for the data lake*: This involves setting up automated CD processes to ensure that the latest versions of the codebase are deployed to the data lake.

The choice of code repository depends on the organizational structure and processes. Generally, the repositories can be with mono-repo or multi-repo. Both have their advantages and disadvantages.

A mono-repo and a multi-repo are two different approaches to managing code in a continuous integration and continuous delivery (CI/CD) process. The choice of which to use depends largely on the size and complexity of the project, but each approach has its own advantages and disadvantages. A mono-repo is a single repository where all code is stored and managed in one place. This approach allows for sharing code easily and simplifying the collaboration process. It also simplifies version control and makes it easier to maintain consistent code across the project. However, it can be difficult to manage a large project in one repository, as there can be a lot of overhead associated with managing a single repository. Additionally, if there are multiple teams working on the project, each team's code can become intertwined, making it difficult to separate out individual contributions.

A multi-repo approach, on the other hand, splits the project into multiple smaller repositories, allowing for easier management. This approach allows teams to work on individual components without having to worry about changes made by other teams. Additionally, it allows teams to easily deploy their code independently of other teams, making the CI/CD process more efficient. However, it can be difficult to keep track of changes across multiple repositories, and there can be overhead associated with managing multiple repositories. Additionally, it can be difficult to share code between repositories. In conclusion, both mono-repo and multi-repo have their pros and cons, and the best approach for a particular project depends on its size and complexity. If the project is small and not heavily interconnected, a mono-repo may be the better option. However, if the project is large and complex, with multiple teams working on different components, a multi-repo approach may be more appropriate. Ultimately, the best approach is to evaluate the project and determine which option is most appropriate.

What we will not cover as part of this chapter are the general concepts of CI/CD (as there are multiple books dedicated to explaining these concepts). I will keep this chapter small and focus on the production path for this customer for the entire data engineering framework that we have created.

Figure 6-1 shows the high-level process we introduced for the code deployment and integration.

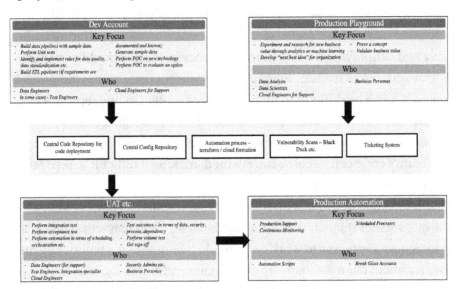

Figure 6-1. *DevOps process for the central framework within the data lake*

The idea is to have separate environments (a lower environment and the production playground that we discussed previously). The environments serve special purposes, as shown in Figure 6-1. The development environment is where most of the data engineering pipelines are built and tested, whereas the playground accounts are where the new business values are evaluated (in terms of data and machine learning products). Once the software pieces are implemented, the idea is to deploy them to production (through a bunch of intermediate environments).

The journey to the production environment is through the automation process, which is highlighted in the figure. This process to produce software pieces from one environment to another is through a code repository for sharing code, a config repository for sharing configs and other environment files, automation deployment scripts like Terraform and Cloud Formation scripts, code scans for security, and ticketing systems for auditability purposes. This process was something our customer uses through the standard practice to build a robust and repeatable deployment process.

Similarly, when we discussed the common framework concepts (in previous chapters), we decided to implement a pod structure for this customer engagement. We divided the teams between a delivery-centric team and a central framework team. The responsibilities of the team were divided so that we did not end up having silos of small frameworks all over the project. The delivery team had responsibility for understanding and documenting the business processes and requirements and implementing them using the common frameworks. They owned the actual pipeline, use cases, and business outcomes. They followed the proper agile process and waterfall model of understanding the delivery scope, implementing the code, unit testing, and moving the deployment to higher environments (including support). The framework pod team was more aligned with the maintenance, extension, customization, and enhancement of the framework as a whole.

The common framework team was dedicated to implementing any new changes needed for any specific use case or any additional new feature that was needed "net-new" because of additional project scope. The intention of this team was to keep the development and maintenance of this framework centralized under one umbrella. Figure 6-2 provides a quick reference of how we managed this. The pod team created new features and built new extensions (based on backlog and prioritization). The framework was tested using synthetic data and went through its own release cycle (including unit test, integration test, and production rollout).

163

Once the new version of the framework was released, it was checked in to the central repository. Multiple project teams (depicted in Figure 6-2) would then take the framework's latest version and use that to build their project-specific pipelines. The pipelines that were created using the latest version of the framework will go through a project-specific testing and deployment cycle.

One additional capability that we introduced for a quick time to market was to enable "user-defined functions" (UDFs) as part of the extension modules. Under that process, the project team can add new code/functions as UDFs only in a specific section of the framework that is called an *extension module*. The extension module is project specific and maintained by the project teams (versus the framework itself that was managed by central pod). We developed a centralized governance process where we evaluated new functions that were added in the extension modules by each project team and then decided if we needed to bring that code back into the core framework so that other projects could benefit from it. That provided a perfect feedback and orchestrated process to ensure the code was managed centrally yet the project teams were not 100 percent dependent on any specific changes that were needed for a project. Figure 6-2 showcases the process flow with Databricks on AWS in mind. The same concept can be applicable for any centralized pod.

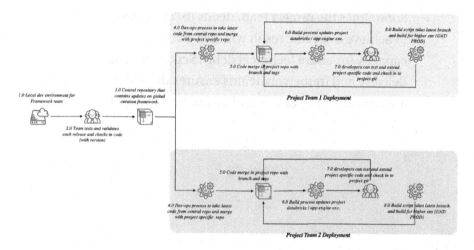

Figure 6-2. *Code/framework release process for cross-domain teams*

Key Takeaways

It is important to separate the delivery pod from the framework pod to ensure a common reusable solution buildout. The following are some of the key advantages for this pattern:

- The framework is centrally managed and does not end up with multiple copies over time.

- It falls nicely within the concept of build once and reuse many times.

- There is a separation of concerns. The project delivery can focus on bringing in business value through the framework without worrying about managing the framework itself.

- For all practical purposes, it is impossible to hire/recruit resources who have deep AWS, PySpark, Python, etc., skills in bulk. A reusable framework helps to reduce the dependency on lots of high-skill resources.

- Sometimes the project team needs to "cut corners" to achieve a specific result within a given time and capacity (or take a technical debt). Separating the delivery and framework teams ensures those decisions are made independently and do not impact other projects.

CHAPTER 7

Miscellaneous

Objective: Advice to Follow

I want to ensure that my experience can add value to your own journey. This last chapter will present some best practices and lessons learned from my experience.

Recommendations

The following sections provide some industry advice based on firsthand experiences and lessons learned.

Managing a Central Framework Along with Project-Specific Extensions

We have talked about multiple frameworks such as the rules engine, data quality engine, data reconciliation engine, and more. All these frameworks help to maintain repeatability and automation. However, managing a central team comes with its own disadvantages. For example, for any small or project-specific changes, the project development team has to raise a request to the central framework team (which will also come with its

© Nayanjyoti Paul 2023
N. Paul, *Practical Implementation of a Data Lake*,
https://doi.org/10.1007/978-1-4842-9735-3_7

own prioritization, etc.). This process creates unnecessary red tape and complications. For this customer, we ensured that we didn't fall into that loop just because we wanted to introduce a centrally managed framework.

For this customer, we created the frameworks in such a way that it was possible to build extensions to the central framework without needing to change the framework itself. In this way, the framework team ships the central core framework and code common to all projects, and the project teams can own extensions, which are plugins to the central framework and contain project-specific functions that are not shared across projects.

Allowing Project Teams to Build "User-Defined Procedures" and Contribute to the Central Framework

We discussed user-defined functions (UDF) as features that project teams can add to a central framework independently without going through the whole nine yards of managing the framework backlog. We also mentioned the extension modules that are managed by the project teams (instead of the central framework team) for delivering project-specific features independently. However, in doing so, we might run the risk of creating too many extension functions by the project teams independently. This might create the same problem where every team ends up creating only extension functions.

To counter that issue, we introduced the framework governance board that has justifications for the new extension functions needed and a cadence for how to bring back code from all extension modules (from different project teams) and put them into the common framework code so that others may benefit. Figure 7-1 provides a detailed process flow on how to set up and manage the framework extension by the project teams through a central audit and governance board.

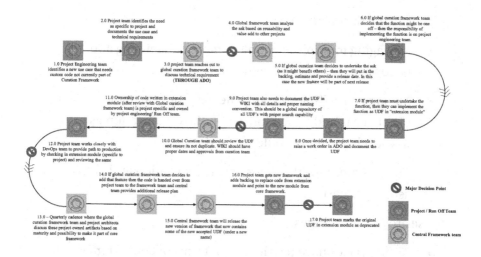

Figure 7-1. *High-level process flow of how to manage the release for each team for a central framework that is owned by central IT*

Here, the central framework team owns the framework itself, but the project team handles the implementation of project-specific requirements via the extension module, which needs to be designed and approved by the central governance board. Once the new extensions are implemented and deployed, frequent revisions and validations are done to check if any of the project-specific extensions can be generalized and moved from the extension module to the core framework. This process also involves governance board process and analysis.

Advantages and Disadvantages of a Single vs. Multi-account Strategy

We talked about the multi-account structure and how the current customer used multiple AWS accounts to set up the production platform. This section talks about the advantages and disadvantages of a multi-account strategy.

These are the disadvantages:

- Managing multiple accounts is more challenging and needs a bigger management team.

- Managing multiple accounts needs proper separation of concerns and education across the organization as to how each account should be used.

- There is more overhead for data classification and security measures.

These are the advantages:

- AWS services have restrictions in terms of concurrency and service limitations. For example, Athena can handle 20 parallel queries; Glue can execute 50 parallel jobs and 1,000 concurrent jobs; and so on. All these limitations need to be managed while executing production-grade pipelines. Separating the production setup logically between multiple AWS accounts helps us solve for the limitations.

- Not everyone and every business process needs all the AWS services and accesses. Creating multiple accounts helps us create different privileges and policies in each of the accounts. Similarly, we discussed the centralized security account. Using such an approach we can handle multiple AWS accounts through a central security account.

- Role- and project-based access control is critical (specifically when we talk about data products). Creating different accounts by organizational units helps to maintain separate access control and security privileges.

Creating a New Organizational Unit AWS Account vs. Onboard Teams to a Central IT Managed AWS Account

Figure 7-2 depicts a simple decision tree on how to decide between onboarding a new tenant (organizational or business unit) to an existing AWS account or onboarding them to a new AWS account.

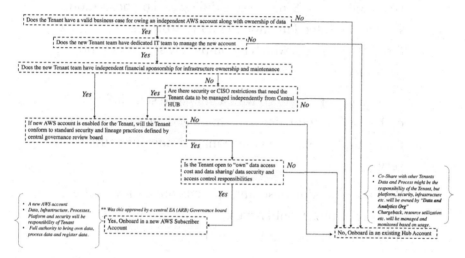

Figure 7-2. *Decision tree for project onboarding into the central data lake*

We followed this process for this customer, and it made things simple for us. The decision is between taking ownership and responsibility independently and relying on the central IT department to run the operations. Once that decision is made, the decision tree can help in deciding if the organizational units can have their own independent AWS accounts in the production setup.

Considerations for Integrating with Schedulers

Orchestration and scheduling become critical when data pipelines are ready to be deployed. For this customer, when we tested multiple data pipelines in a development environment and were ready to schedule and orchestrate them in a higher environment, we had to decide which one to choose from. We had couple of options.

- We could use AWS-native scheduling and orchestration services like Step Function (with EventBridge, CloudWatch, etc.), which is completely serverless and AWS specific.

- We had the option to go with a managed scheduling and orchestration setup like AWS-managed airflow. This was costly and was more on the managed side rather than serverless.

- We had the option of setting up a custom airflow in the EC2 box, which would have been cheaper but needed more maintenance and administration.

- We could use some third-party scheduler and orchestration tool like control-M, which is for enterprises and licensed.

Now, every customer and their requirements might need us to make a different decision, but for this customer we chose the control-M option, as it was an already existing setup. All their on-premise processes were managed by control-M, and it was integrated with a ticketing and notification system. However, all the previous options came with their own advantages and disadvantages, and it is important to evaluate each one before making a decision.

Choosing a Data Warehouse Technology

Possibly the most important decision in terms of the technology selection is the data warehouse. Typically the Gold layer (Provisioned zone) is associated with a data warehouse (in a lake house architecture). This decision is both critical and difficult as all the cloud vendors have comparable data warehouse solutions and some vendors specialize in data warehouses (like Snowflake, etc.). As technical architects, when we are tasked with the recommendation of choosing a data warehouse platform, we must not choose based on the "here and now" but based on "future readiness" and "commitment" from the players.

There are no right or wrong answers in this section, but you must consider the following points while making a decision:

- Do not think of a data warehouse solution independently; think of it based on the data platform and how the data warehouse needs to interact with other pieces (like the Curated zone, data obfuscation, data catalog, etc.).

- Evaluate the data warehouses personally. Most of the details on the Internet are stale (considering the speed of innovation and new features added by the vendors).

- Consider the capabilities offered by vendors in terms of decoupled data from storage. This is critical as the data volume grows and as more business units are onboarded.

- Check on the performance numbers and matrices. When things execute in production, SLAs become the most critical metric.

- Check the data sharing capabilities and features offered by the vendors. With the current trends of decentralized data products and ownership, this feature becomes critical.

- Decide if a multicloud is critical and important for the customer, and if so, choose a data warehouse that can be agnostic to a cloud vendor.

Managing Autoscaling

We have been talking about cloud-centric data platform solutions centered around serverless and managed services. These new SaaS offerings from the cloud vendors ensure shared responsibilities and easy management and maintenance for large-scale production setups. However, there are a few gotchas and caveats that we need to be aware of when it comes to autoscaling (and we should architect and design solutions around these limitations). Table 7-1 provides some of the lessons learned over the course of this book.

Table 7-1. *Learned Considerations for selecting AWS services for managing Auto Scaling*

Service Name	Lessons Learned
Lambda	Lambda will run within VPC. Lambda will be used by the Orchestration service, UI services, and back-end services.
AWS Glue	AWS Glue should run within VPC. AWS Glue needs access to other AWS services (like S3, Redshift) through VPC Endpoint. 50 jobs can run in parallel (soft limit). 1,000 concurrent job runs per job (soft limit). 1,000 jobs per account (soft limit).
S3	100 buckets per account. Enable a VPC endpoint to specific buckets (like Gold and Silver).
Athena	20 queries in parallel. 30 minutes per query.
Redshift	How to handle auto increase of volume (elastic resize). Concurrent scaling of user queries (concurrency scaling for WLM queue). Multiregion data share not available.

Managing Disaster Recovery

One of the advantages of being in the cloud is that it provides high availability and high resiliency. However, there are a few caveats. Cloud vendors such as AWS provide cross-availability, not cross-region, zone recoverability. This means a failure to one of the data centers in US East (like Ohio) will not impact the data, as by default another availability zone (like N Virginia) that belongs to the same US East region always has a copy of the data. However, this setup does not guarantee a failure of all data centers in the US East region, and as there is no default copy of the data in another region (like US West), the customer has a risk of losing his data. To solve this issue, we can design a disaster recovery (DR) solution, which provides cross-region (or in extreme cases multicloud) backups.

Backups in the cloud follow the same principle. We can enable DR to be HOT-HOT, which means the same copy of data is always loaded and processed in two different regions independently so that failure to one will not impact the other. This setup is the most complex and expensive solution to DR. The other option is HOT-WARM, which means the other region contains the same data, but instead of processing the same data in both the regions, typically the data is processed in one region and sent to another immediately. This is a happy medium ground where the data is made available as soon as possible, but it is not processed twice. The third option is HOT-COLD; as the name suggests, the data is synced up once a day or week to ensure we can fall back to a common baseline in case of failure. This solution is the most cost effective. However, this solution does not guarantee 100 percent of data recoverability and can have the issue of data loss.

In addition to the three options to enable a DR strategy, concepts like recovery point objective (RPO) and recovery time objective (RTO) provide the organization with information about which datasets and processes are critical and which are not. This can help provide some guidance to make some sections of the data platform HOT-HOT, whereas other sections (and business processes) can be enabled with HOT-COLD, etc.

AWS Accounts Used for Delivery

We talked about the multi-account strategy; however, in this section we will just provide Table 7-2, which lists the kinds of accounts we enabled for the customer.

Table 7-2. *Mutli-account Strategy*

AWS Accounts	Description	Primary Roles	Access Patterns
Development	No production data and used by developers to test code.	Data engineers, cloud engineers	Sample data read and write into the Landing, Raw, and Provisioned (PDP) areas using automation and scheduling
Staging automation	Staging account for sample production data for testing integration, performance, business acceptance	Data engineers, cloud engineers	Production data (samples) read and write into the Landing, Raw, and PDP areas using automation and scheduling
Staging playground	Staging account for exploratory work	Data scientists, data analysts, functional analysts, data engineers	Read-only cross-account (staging automation) access to Raw and PDP. Additional write access to HOME and Shared folders within S3 of the staging playground.
Production automation	Account for all production-deployed jobs and data.	Data engineers, cloud engineers	Production data read and write into Landing, Raw, and PDP areas using automation and scheduling

(continued)

Table 7-2. (*continued*)

AWS Accounts	Description	Primary Roles	Access Patterns
Production playground	Production account for only exploratory work	Data scientists, data analysts, functional analysts, data engineers	Read-only cross-account (production automation) access to Raw and PDP. Additional write access to HOME and Shared folders within S3 of the production playground.

Data Platform Cost Controls

A data lake provides a decentralized and democratized analytical capability for business units and domains within the organizations to provide business value through data. So when things run in production, it becomes important to evaluate the cost of managing and maintaining the data platform as a whole. Within the construct of data lakes, the cost controls can be categorized into five buckets (Figure 7-3).

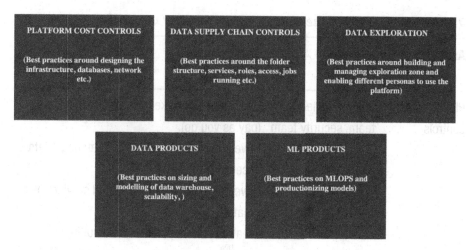

PLATFORM COST CONTROLS

(Best practices around designing the infrastructure, databases, network etc.)

DATA SUPPLY CHAIN CONTROLS

(Best practices around the folder structure, services, roles, access, jobs running etc.)

DATA EXPLORATION

(Best practices around building and managing exploration zone and enabling different personas to use the platform)

DATA PRODUCTS

(Best practices on sizing and modelling of data warehouse, scalability,)

ML PRODUCTS

(Best practices on MLOPS and productionizing models)

Figure 7-3. High-level cost drivers for a data lake

These cost controls provide a good reference to understand the cost and ownership distribution. Additionally, they can help manage the budget and plans for investment based on the outcomes from these cost controls. Table 7-3 provides a good reference to the cost controls.

Table 7-3. *Details of the Cost Controls for the Data Lake*

Area	Impacted Personas	Cost Controls
Platform cost controls	Cloud engineering team, security team	Use serverless to keep control of the costs (pay as you go). Have a dedicated CMK KMS to manage data access. Have appropriate user roles and service roles created. Break down the platform into multiple cloud accounts. Have a dedicated VPC/VNET and divide the platforms into multiple subnets. Enable proper network guardrails to ensure connections between accounts. Have cost monitors enabled to monitor usage.
Data supply chain cost controls	Cloud engineering team, data engineering team	Have dedicated service roles. Enable different kinds of data engineering roles to access separate service roles. Enable guardrails on the data pipeline in terms of resources, cost, runtime, data processed, etc. Enable only required resources. Have specific "break-the-glass account" for data pipelines instead of giving everyone access to the pipelines.

(*continued*)

Table 7-3. (*continued*)

Area	Impacted Personas	Cost Controls
Data exploration cost control	Cloud engineering team, data science team, data analyst team	Enable the concept of shared versus restricted. Have role- and attribute-based controls as to what roles can access what section of data. Enable controls over data access, data sharing, collaboration, etc. Have a cost monitor to ensure the monitoring of resources usage. Invest in training and building common shared artifacts. Have a dedicated cadence on cleanup and review.

Table 7-4. *Details of the Cost Controls and Their Implications for Each Team*

Area	Impacted Personas	Cost Controls
Data product cost controls	Cloud engineering team, data engineering team	Use serverless to keep control of costs (pay as you go). Identify the right-fit technology for each data product. Have specific roles and usage patterns for users of the data products. Invest in the right data model based on bottom-up approach. Leverage decoupling of compute versus data to keep users and data separate.
ML product cost control	Cloud engineering team data science team data engineering team	Focus on serverless to build and deploy the end-to-end MLOPS process and pay for use only. Enable separate roles for training versus deployment. Restrict resources for a training account. Enable/disable certain capabilities of MLOps based on account (like prod versus playground). Define and establish a proper CI/CD control.

Table 7-5. *Details of The Cost Controls and the Areas They Impact*

Area	Section	Cost Controls
Data supply chain	Platform and resources	Use of serverless for "pay as you go" and focus on cloud native versus bring your own. Use of serverless to scale resources on demand based on workloads. Have controls on lower and upper limits of resources that can be allocated using IAM. Allow lower and upper bounds on GPUs for Glue, EMR, etc. Allow upper and lower bounds of memory restrictions on Lambda. Have upper- and lower-bound resources around autoscaling for RDS, etc. Have a proper grooming process for data onboarding with spec/config file to validate before onboarding datasets to avoid loading tons of data. Restrict number of reruns, etc., based on error code and error condition. Use DynamoDB or other on-demand data storage specifically to store configs, spec, audits, etc., with on-demand scaling. Enable timeouts for long-running ETL. Enable TTL for objects that are needed for a specific time.

(continued)

Table 7-5. (*continued*)

Area	Section	Cost Controls
Data exploration	Buckets and structures	Each user should have a dedicated HOME directory, and every object created in a user's HOME should be tagged for cost control. There should be a dedicated SHARED directory with PUBLIC, ROLE based, and PROJECT BASED subfolders with dedicated owners of each directory for cost controls. Automated process to have weekly report to clean/promote artifacts from exploration to avoid making exploration shadow IT.
Data exploration	Ad hoc work	Define dedicated workgroups and align users to specific workgroups. Have resource limits on the number of concurrent queries, amount of data that can be scanned, etc. Set the max query timeout to avoid long-running queries. Tag each query against the profile to have reports around the number of queries and the amount of data scanned per query for each user. Restrictions on how large a dataset someone can bring into the HOME directory for exploration. Restrictions on data download capability in exploration zone.

(*continued*)

Table 7-5. (*continued*)

Area	Section	Cost Controls
	Resources and features	Add a specific service per role; not everyone should have access to every service
		Add tags to every process created in the exploration zone such as Glue jobs, etc.
		Disable scheduling and regularly review and promote/delete the jobs.
		Limit concurrent jobs in exploration, resources allocated to jobs in exploration.
		Provide options to users to add tables in the exploration zone manually by running templates instead of running scheduled crawlers, etc.
		Have dedicated roles for each data scientist to work on dedicated resources for ML modeling (like Sagemaker, etc.).
		Enable life-cycle policies to shut down resources based on idle time.
		Use policies to restrict the number of machines that can be requested/used for testing ML models.
		Have policies around containers as to how many can be launched and the max time before shutdown.
		Restrict processes in ML around real-time endpoint, etc.
		Restrict the user to create a resource in VPC only.

(*continued*)

Table 7-5. (*continued*)

Area	Section	Cost Controls
Data products	Building fit for purpose data models	Choice of data warehouses that can decouple storage from compute so that one can scale irrespective of the other.
		Use concepts of zero-copy to have different organizational units use same centralized dataset.
		Have restrictions on the compute for each role. For example, data scientists only have access to smaller compute versus automated processes.
		Use autoscaling to leverage cloud capability but always enable lower and upper bounds on resources.
		Have monitoring of daily/weekly usage and change the autoscaling policy over time (do not keep it static).
ML products	MLOps	Have a standard CI/CD pipeline to allow data scientist requesting models to be deployed to production (without giving free reign to the data scientist).
		Have a business decision around the following:
		How frequently the data characteristics change so that model has to be re-trained (avoid frequent retraining).
		Classify the model importance and decide how often to rebuild models and hand them off to production.
		Categorize models into batches and stream and solve accordingly (do not build everything).
		Have a dedicated "break-the-glass" account in production for ML jobs without providing every data scientist access to the Production account.
		If the model updates/inserts a database/warehouse, then design the warehouse accordingly with optimal resources.

Common Anti-patterns to Avoid

Avoid the following common anti-patterns.

One-Size-Fits-All

Avoid: Designing a data lake to fit all types of data without understanding the specific data requirements

Designing a one-size-fits-all data lake is not a good idea because it may not be able to accommodate the needs of all users. Each user may have different requirements that need to be addressed, and a one-size-fits-all data lake may not be able to fulfill all of them. Additionally, it may be difficult to scale the data lake to accommodate different use cases and ever-changing data requirements. Finally, it may be difficult to maintain and secure the data lake if it is not optimized for the specific use case. Hence, it is important to separate the data lake into zones and have specific data storage, access, and use case patterns for each zone separately.

Ignoring Security

Avoid: Failing to implement proper security measures, such as authentication and authorization, to prevent data breaches

Setting up proper security measures is essential to prevent data breaches because it helps to protect the confidential and sensitive data that businesses and organizations have stored. Authentication and authorization provide a means to verify the identity of users who are attempting to access the data and restrict access to only those who are authorized. This helps to ensure that only those with the proper credentials can access the data and helps to prevent unauthorized access that can lead to data breaches.

Data Sprawl

Avoid: Not keeping track of the data stored in the data lake, leading to data duplication and lack of clarity

Data sprawl is the uncontrolled growth of data across an organization's systems. It occurs when data is stored in multiple systems or when systems become overloaded with data. This makes it difficult to access, manage, and analyze the data, and it can lead to increased storage and maintenance costs. Data sprawl in data lakes should be avoided to ensure that data remains organized, secure, and accessible. Data sprawl can lead to data duplication, data inconsistency, and difficulty in managing and maintaining the data. Additionally, it can lead to security risks as unstructured data is often not as secure as structured data. Finally, it can lead to performance issues because of the sheer volume of data that needs to be managed.

Poor Data Governance

Avoid: Not having clearly defined roles and responsibilities for the data lake, leading to data inconsistency

Poor data governance can lead to the following:

- *Quality issues:* Poor data governance can result in a data lake filled with low-quality data that is inaccurate, incomplete, or out-of-date. This can lead to inaccurate insights, which will in turn lead to poor decision-making.

- *Compliance and regulatory issues:* Poor data governance can lead to data lakes that are not compliant with industry regulations, leading to financial and legal repercussions.

- *Security issues*: Poor data governance can result in data lakes that are vulnerable to data breaches and other security issues.

- *Lack of trust:* Poor data governance can lead to a lack of trust in the data lake, resulting in users being hesitant to use the data lake and resulting in decreased adoption and usage. Hence, our data lake implementation should have a "governance-first" approach and not be an afterthought.

Lack of Quality Controls

Avoid: Not having a system in place to ensure data accuracy and reliability

The lack of quality controls can have a huge impact on data lakes. Poor-quality data can lead to inaccurate results and poor decision-making. It can also lead to data inconsistency and corruption, making it difficult to get meaningful insights from the data. It can also lead to increased costs as it may take longer to clean up the data and make sure it is accurate. Additionally, it may lead to the wrong decisions being made based on the data, which can have long-term consequences.

Poor Metadata Management

Avoid: Not tracking the data stored in the data lake, leading to data duplication and lack of clarity

This can lead to the following:

- *Difficulty in reuse:* Poor metadata management can make it difficult to locate data that could be reused in another project or application. Without an organized and well-structured metadata repository, users may struggle to discover what data is available and how it can be used.

- *Security risks:* Poor metadata management can also leave data lakes vulnerable to security risks. Metadata can be used to track who is accessing data and when it is accessed. If this information isn't properly managed, it can leave the data lake open to malicious actors who may be able to access sensitive data.

- *Lack of data governance:* Poor metadata management can also lead to a lack of data governance. Without an organized and well-structured metadata repository, it can be difficult to establish data governance policies and ensure compliance. This can leave the data lake open to potential misuse and abuse.

- *Reduced performance:* Poor metadata management can also lead to reduced performance. If a data lake is not properly indexed and managed, it can take longer for users to find the data they need. This can lead to delays in data analysis and decision-making, resulting in a reduced ROI.

Wrong Tools

Avoid: Using the wrong tools and technologies for the data lake implementation, leading to inefficient results

Tool and vendor selection is critical for successful data lake implementation as it contributes to the success of the overall project. The right tool and vendor will provide the necessary capabilities, scalability, and support to ensure that the data lake can meet the organization's requirements. The wrong tool and vendor can result in costly delays or data loss, or even a complete failure of the implementation. The selection process should consider the following criteria: cost, scalability, data governance, security, reliability, performance, integration, and extensibility. The right tool and vendor will help maximize the value of the data lake by providing the necessary features and capabilities to meet the organization's data requirements.

Avoid Over-Engineering

Avoid: Making the data lake too complex for the problem it is supposed to solve

This leads to the following:

- Using an overly complex and expensive technology stack to build the data lake, such as using an enterprise-level data warehouse solution when a much simpler and cheaper solution would work.

- Including a large number of data sources when a few key sources are sufficient.

- Over-normalizing data into numerous tables and columns when a single table with a few columns of data would suffice.

- Building multiple pipelines for the same data source when one pipeline could do the job.

- Overloading the data lake with nonessential data that may not be used in the future. Building a complex data governance framework for the data lake when a simpler one could work.

Poor Data Integration

Avoid: Not having a data integration approach in place to ensure data consistency and accuracy

Poor data integrations can lead to inaccurate or incomplete data lakes. If the data lake does not contain all of the necessary data, then it cannot be used to its full potential. Additionally, if the data integrations are not done correctly, the data lake may contain data that is inaccurate or not up-to-date, leading to incorrect or incomplete analysis. Furthermore, poor data integrations can lead to significant delays in the data lake's development, as well as an increase in operational costs. Finally, poor data integrations can lead to data lakes that are difficult to maintain, as errors and discrepancies need to be manually identified and corrected.

Unstructured Data Overload

Avoid: Storing too much unstructured data in the data

This can lead to the following:

- *Difficulty in searching data*: With the unstructured data overload, it can be difficult to find the right data that you're looking for. This can be especially challenging in data lakes, which are often vast repositories of data.

- *Time-consuming ETL processes*: Extracting, transforming, and loading (ETL) data from the data lake can be a very time-consuming process. This is because of the sheer volume of data that needs to be processed, as well as the complexity of the data.

- *Lack of data validation*: With unstructured data, there is often no way to validate the data to ensure that it is accurate or complete. This can lead to unreliable data, which can have serious implications for decision-making.

- *Security risks*: Unstructured data overload can lead to security risks, as unstructured data may not be subject to the same rigorous security standards as structured data.

- *Data governance and compliance issues*: With unstructured data overload, it can be difficult to ensure that the data is compliant with the organization's data governance policies and regulations. This can lead to compliance issues, which can have serious consequences.

Key Takeaways

We have come to a logical end to this project implementation. Thanks for sticking around for my story. This is where we did all the hard parts and ensured that we have logically and technically enabled each piece of the "data strategy puzzle." I must admit that I was overwhelmed at first, but when I thought of solving things in their own logical sequence and focused on one step at a time, things became simple and possible. I hope this closing chapter has provided additional insight and solutions for any issues you may encounter.

Index

A

Ability, 47, 109, 139, 156
Access control process, AWS Service
    Athena, 39
    CloudWatch, 41
    DVL (EC2), 40
    DynamoDB, 42
    EMR (Spark), 38
    Glue, Glue Crawler, 42
    lambda functions, 35
    Redshift, 39
    S3, 32
    Sagemaker, 38
    SNS, 38
    step functions, 35
Account strategy, 8
Active Directory (AD)
    integration, 31
Advanced analytics user, 156
Alert monitoring systems, 145
Analytics platform, 2, 46, 50
Assessment questions, 4
Athena, 23, 138, 175
Authentication, 25, 31, 187
Authorization, 31, 187
Automation, 107, 140, 151, 158,
    160, 163

Autoscaling, 174, 175
AWS accounts, 169, *See also*
    Multi-account strategy
AWS Glue, 138, 175
AWS Key Management
    System (KMS)
    AMK, 74
    AOK, 74
    asymmetric keys, 74
    CMK, 74
    data keys, 74
    encrypt/decrypt data, 73
    encryption and decryption
        process, 74
    symmetric keys, 74
AWS Lake Formation, 22, 25,
    137, 140
AWS-managed airflow, 172
AWS Managed Keys (AMK), 74
AWS Owned Keys (AOK), 74
AWS services, 44, 145, 170

B

Blast radius, 24, 47, 50, 68, 73
Business-changing ideas, 151
Business intelligence (BI), 115, 132
Business metadata, 101

Printed in the United States
by Baker & Taylor Publisher Services